NATIONAL DEFENSE RESEARCH INSTITUTE

T0131045

Ethics in Scientific Research

An Examination of Ethical Principles and Emerging Topics

Cortney Weinbaum, Eric Landree, Marjory S. Blumenthal,
Tepring Piquado, Carlos Ignacio Gutierrez

Prepared for the Intelligence Advanced Research Projects Activity

For more information on this publication, visit www.rand.org/t/RR2912

Library of Congress Cataloging-in-Publication Data is available for this publication.
ISBN: 978-1-9774-0269-1

Published by the RAND Corporation, Santa Monica, Calif.
© Copyright 2019 RAND Corporation
RAND® is a registered trademark.

Support RAND
Make a tax-deductible charitable contribution at
www.rand.org/giving/contribute

www.rand.org

Preface

The goal of this project was to provide researchers, government officials, and others who create, modify, and enforce ethics in scientific research around the world with an understanding of how ethics are created, monitored, and enforced across scientific disciplines and across international borders. The research had two motivations: (1) to inform researchers and sponsors who engage in research in emerging scientific disciplines and who may face new ethical challenges, and (2) to inform research sponsors—including government officials—who wish to encourage ethical research without unintentionally encouraging researchers to pursue their research in other jurisdictions.

While this report is intended for audiences with scientific expertise, it is also intended to be useful to scientists from various disciplines who have differing expertise from other disciplines and for citizen scientists and amateur researchers who lack formal scientific training. Therefore, it should be accessible to informed readers from a variety of backgrounds and experiences.

This research was sponsored by the Intelligence Advanced Research Projects Activity (IARPA) and conducted within the Cyber and Intelligence Policy Center of the RAND National Defense Research Institute, a federally funded research and development center (FFRDC) sponsored by the Office of the Secretary of Defense, the Joint Staff, the Unified Combatant Commands, the Navy, the Marine Corps, the defense agencies, and the defense Intelligence Community. All of the data collection and analysis described in this report took place in 2018.

For more information on the RAND Cyber and Intelligence Policy Center, see www.rand.org/nsrd/ndri/centers/intel or contact the center director (contact information is provided on the webpage).

Contents

Figure, Tables, and Boxes

Figure

Tables

Boxes

Summary

Scientific research ethics vary by discipline and by country, and this study sought to understand those variations. Our team reviewed literature from across scientific disciplines and conducted interviews with experts in the United States, Europe, and China. Our analysis led to an understanding of which ethics are common across disciplines, how these ethics might vary geographically, and how emerging topics are shaping future ethics. We focused on the ethics of scientific research and how the research is conducted, rather than on how the research is applied. This distinction excluded from our research an analysis of so-called "dual use" applications for military purposes.

Our literature review of more than 200 documents led us to identify ten ethical principles that are generally common from one scientific discipline to another, shown in Table S.1.

We found that these principles often can be traced to such foundational documents as the Nuremberg Code (in 1947), the Universal Declaration of Human Rights (1948), the Declaration of Helsinki (1968), the European Charter of Fundamental Rights (2000), and others. Our research found that ethics are created, change, and evolve due to significant historic events that create a reckoning (e.g., the Nuremberg trials), due to ethical lapses that lead researchers to create new safeguards (e.g., the Tuskegee Study), due to scientific advancements that lead to new fields of research (e.g., the emergence of experimental psychology), or in response to changes in cultural values and behavioral norms that evolve over time (e.g., perceptions of privacy and confidentiality).

We found that instances of an ethical principle from Table S.1 not being discussed in a particular discipline's code of conduct can often be attributed to that discipline not engaging in research on human or animal subjects. Ethics such as informed consent and beneficence—which relate to the informed consent *of* the human subject or the beneficence *toward* the human or animal subject—are missing from disciplines without research participants, including most physical sciences and mathematics. This led to our next finding, that certain ethical principles are common across all fields, while other principles are specific to certain types of research, as shown in Table S.2. This finding could help researchers in a new emerging discipline determine which ethical principles best apply to their work.

Table S.1
Ethical Principles for Scientific Research

Ethical Principle	Definition
Duty to society	Researchers and research must contribute to the well-being of society.
Beneficence	Researchers should have the welfare of the research participant in mind as a goal and strive for the benefits of the research to outweigh the risks.
Conflict of interest	Researchers should minimize financial and other influences on their research and on research participants that could bias research results. Conflict of interest is more frequently directed at the researcher, but it may also involve the research participants if they are provided with a financial or nonfinancial incentive to participate.
Informed consent	All research participants must voluntarily agree to participate in research, without pressure from financial gain or other coercion, and their agreement must include an understanding of the research and its risks. When participants are unable to consent or when vulnerable groups are involved in research, specific actions must be taken by researchers and their institutions to protect the participants.
Integrity	Researchers should demonstrate honesty and truthfulness. They should not fabricate data, falsify results, or omit relevant data. They should report findings fully, minimize or eliminate bias in their methods, and disclose underlying assumptions.
Nondiscrimination	Researchers should minimize attempts to reduce the benefits of research on specific groups and to deny benefits from other groups.
Nonexploitation	Researchers should not exploit or take unfair advantage of research participants.
Privacy and confidentiality	Privacy: Research participants have the right to control access to their personal information and to their bodies in the collection of biological specimens. Participants may control how others see, touch, or obtain their information. Confidentiality: Researchers will protect the private information provided by participants from release. Confidentiality is an extension of the concept of privacy; it refers to the participant's understanding of, and agreement to, the ways identifiable information will be stored and shared.
Professional competence	Researchers should engage only in work that they are qualified to perform, while also participating in training and betterment programs with the intent of improving their skill sets. This concept includes how researchers choose research methods, statistical methods, and sample sizes that are appropriate and would not cause misleading results.
Professional discipline	Researchers should engage in ethical research and help other researchers engage in ethical research by promulgating ethical behaviors through practice, publishing and communicating, mentoring and teaching, and other activities.

NOTE: *Research participant* refers to someone with an active role participating in research, whereas *research subject* could include someone whose data are used but who does not consent to participate.

When we examined ethics by geographic region, we found a distinction between the ethical principles of researchers conducting research in their own host country compared with the ethical principles of researchers conducting research in a country *other* than their native country. This distinction is important in understanding whether ethical differences are a result of local customs, culture, laws, and practices or instead result from one culture being subjected to the ethics of foreigners. In either

Table S.2
Categories of Ethical Principles

Category	Description of Category	Ethical Principle
Ethical scientific inquiry	The research inquiry itself must benefit society.	• Duty to society
Ethical conduct and behaviors of researchers	Researchers should conduct themselves in certain manners, and they are responsible for their knowledge and awareness of ethics and appropriate research methods.	• Conflict of interest • Integrity • Nondiscrimination • Professional competence • Professional discipline
Ethical treatment of research participants	Research participants should be treated according to certain guidelines and treated humanely, and the environmental or secondary effects of the research should be considered.	• Informed consent • Beneficence • Nondiscrimination • Nonexploitation • Privacy and confidentiality

situation, we found that the Declaration of Helsinki, which was written specifically for physicians, provides a common ethical foundation for any researcher anywhere in the world, regardless of whether his or her country has well-developed ethical standards, monitoring, or enforcement mechanisms.

In situations in which researchers conduct research outside their home country, we found concerns of *ethics dumping*, a term coined to describe when a researcher conducts research in a region with less stringent ethical requirements than his or her own homeland. Obviously, this creates its own ethical dilemma (e.g., how can avoiding ethics be ethical?), yet several scientific disciplines require foreign travel to collect genetic data from or conduct anthropological research on indigenous peoples. Several governments, including the European Commission, have created guidelines for research conducted with their funds in foreign lands to mitigate the negative effects of such research on local populations. One indigenous group, the San people in Africa, has even created its own code of conduct, which the tribe published in English and requires all researchers to agree to before beginning research on its members.

As we examined laws, regulations, and standards around the world, we sought an understanding of the relationship between enforceable laws and unenforceable norms. We undertook this project with an understanding that laws can be unethical and ethics can be unregulated, and we hoped to understand how the administration of both laws and ethics may better align. We found that professional societies and journals aim to fill the gap between laws and ethics by documenting ethics that they expect of their members or authors, respectively; requiring members and authors to self-certify that they complied with such rules; and providing a reporting or grievance mechanism for cases in which members or authors self-report or are reported on. When these societies or journals have international membership or readership, they additionally seek to smooth out ethical differences from region to region by creating a discipline-wide standard. These mechanisms are imperfect, however, as they are not enforceable and often not even monitored.

Here, we found that emerging topics in ethics may both help and strain this current arrangement. As open science continues to give more researchers access to each other's data and results, researchers may find their work more often peer reviewed by a wider audience (thanks to open data) or less frequently peer reviewed (thanks to open access publication). Transparency is an ally of ethics, yet the open science movement creates its own challenges. Meanwhile, citizen science, the movement of more amateur scientists participating in research, opens the doors to wider societal participation in scientific advancements along with the risk of a greater number of researchers who lack formal training in ethics or in choosing appropriate research methods. We found that society as a whole, and professional scientific societies specifically, could take proactive steps to engage citizen scientists in dialogues about research ethics. Today, this is particularly needed in topics that include big data and biology, two areas where citizen scientists may test the limits of research ethics.

In these disciplines and others, we found a new emerging discussion in literature on *bystander risk*, the risk levied on people who did not consent to participate in research because they are unintended bystanders. In medical research, this may include a person who comes in contact with a research participant who has been exposed to a contagious disease; in this case the research participant consented, but the persons exposed did not consent. In information science research, bystander risk may include the persons whose privacy is compromised due to big data analytics, including analytics on data they themselves did not create nor authorize for use.

Finally, across our research, we found several pillars that researchers and sponsors can lean on to promote and strengthen ethics—specifically, education and training of ethics and research methods; professional societies and communities that promulgate and advocate codes of ethics; and governance mechanisms that range from institutional oversight to law and regulation.

Acknowledgments

This project would not have been possible without sponsorship from the Intelligence Advanced Research Projects Activity (IARPA). We thank Jason Matheny, former director of IARPA, for having a vision of how IARPA could contribute to a societal dialogue about ethics in emerging scientific disciplines, and we are grateful that Stacey Dixon, current director of IARPA, and Jeff Alstott, our project monitor, share Jason's vision, helped us see it through, and provided us course corrections whenever our project veered astray.

This study was a team effort, and the team was significantly larger than just the authors. We thank our RAND colleagues (in alphabetical order) Nathan Beauchamp-Mustafaga, Amanda Edelman, Betsy Hammes, Scott Harold, Raza Khan, Gordon Lee, Alexis Levedahl, Alice Shih, Jirka Taylor, and Keren Zhu for helping us gather and code the literature we analyzed, translating foreign-language documents into English and translating our interview protocol into Mandarin Chinese, helping us identify and connect with potential interview participants, editing and reviewing this report, and various other activities that improved the quality of our final work. We thank our RAND Europe colleagues, and specifically Salil Gunashekar, for helping to shape the direction of our international research by introducing us to organizations and experts whom we might not otherwise have found.

Our research would not have been complete without the participation of experts around the world who took part in our interviews. They shared their time and insights with us, and we thank them. Each person's official title and organizational affiliation is included in Appendix D. In alphabetical order, we thank Jeremy Berg, Stephanie J. Bird, Diana Bowman, Raja Chatila, 从亚丽Yali Cong, C. K. Gunsalus, Andrew M. Hebbeler, Rachelle Hollander, Elsa Kania, Isidoros Karatzas, Annie Kersting, Gary Marchant, Anne Petersen, Edward You, and others who requested that we not include their names. You know who you are, and we appreciate your participation in this research.

Thank you to Rebecca Kukla at the Georgetown University Kennedy Institute of Ethics and Ritika Chaturvedi at RAND for identifying gaps in our analysis and areas where our work could be improved. We appreciate the detailed review that you provided.

Nearly a year before this project was funded, we challenged a group of students at the Maxwell School of Citizenship and Public Affairs at Syracuse University to research and review existing codes of conduct for lessons learned that could be applied to new emerging disciplines. They accepted our challenge enthusiastically, and their work informed our research methodology. Thank you to Kashaf Ud Duja Ali, Eni Maho, Earl W. Shank, Derrick J. Taylor, and their advisor, Renée de Nevers.

Throughout this research, we learned that ethics will evolve continuously over time, and we hope this report provides a positive contribution to the international dialogue. We could not have done so without the help of each of you.

Abbreviations

AI	artificial intelligence
ACM	Association for Computing Machinery
AMA	American Medical Association
APA	American Psychological Association
API	application program interface
ASM	American Society for Microbiology
CBD	Convention on Biological Diversity
EC	European Commission
EU	European Union
FDA	U.S. Food and Drug Administration
GDPR	General Data Protection Regulation
GINA	Genetic Information Nondiscrimination Act
HHS	U.S. Department of Health and Human Services
IARPA	Intelligence Advanced Research Projects Activity
IEEE	Institute of Electrical and Electronics Engineers
IRB	Institutional Review Board
ISA	International Sociological Association
IVSA	International Visual Sociology Association

OECD	Organisation for Economic Co-operation and Development
OHRP	Office for Human Research Protections (U.S. Department of Health and Human Services)
R&D	research and development
UN	United Nations
WMA	World Medical Association

CHAPTER ONE

Introduction

Why We Did This Research

The goal of this project was to provide researchers, government officials, and others who create, modify, and enforce ethics in scientific research around the world with an understanding of how ethics are created, monitored, and enforced across scientific disciplines and across international borders. The research had two motivations: (1) to inform researchers and sponsors who engage in research in emerging scientific disciplines and who may face new ethical challenges, and (2) to inform research sponsors—including government officials—who wish to encourage ethical research without unintentionally encouraging researchers to pursue their research in other jurisdictions.

This project focused on the ethics of scientific research. We sought lessons for how researchers could conduct research in an ethical manner in fields such as artificial intelligence (AI) and neurotechnology regardless of whether the findings from the research might be applied to beneficent goals (e.g., to assist persons with disabilities in leading fulfilling lives) or harmful goals (e.g., increasing the severity of warfare). Because so many outputs of science and engineering research can be used in civilian or military contexts (so-called *dual use*), our goal was to understand the ethics that shape how research is designed, conducted, and disseminated independent of how it might be used—which might not be known until long after the research is completed. We sought to describe common ethical principles and differences that exist across scientific and technical disciplines and emphases.

Research Methodology

The majority of our analysis was based on reviewing secondary sources (mostly journal articles) and consulting experts. We started by examining literature, and where we found gaps—questions the literature did not answer—we sought additional commentary and documents and conducted interviews with experts. Our team collected the 200 most-cited peer-reviewed articles on ethics in scientific research from literature over time plus codes of ethics from across scientific disciplines. We began by searching Scopus and Web of Science for the most-cited articles about ethics and research or codes of conduct and research. All of the articles we reviewed had been cited more than 300 times by other authors; the most-cited article had been cited over 5,700 times at the time of our data collection. We chose to focus on the most-cited articles to examine topics that researchers and scientists themselves have indicated are worthwhile. We then searched for codes of conduct from every scientific discipline in which we had literature, and we tagged or coded all of these documents by topic. We analyzed the results for commonalities and differences across scientific disciplines. Where we found gaps in the literature, we conducted interviews with experts in relevant fields, and we searched for additional relevant documents outside of our initial scope of 200 peer-reviewed journal articles.

Our methodology includes several limitations, including that citations are a lagging indicator of how useful researchers found the article, highly cited articles may have been controversial rather than widely accepted, and the articles we collected may reflect the topics that were in vogue at their time of publication rather than more-recent topics. Appendix A includes a detailed description of our methodology, including how we chose which articles to review and which experts to interview, and how we conducted our interviews and our analysis. We discuss this and other limitations of our methodology in Appendix A.

Appendix B includes the codebook we used to code the documents; an explanation of how we created this codebook is provided within Appendix A. Appendix C provides the informed consent and interview protocol we used for our interviews, and Appendix D provides a list of our interview participants, with their titles and organizational affiliations. All interviews were conducted under the agreement that we would not quote interview participants; therefore, we provide this list of names without citing them directly within the report. Appendixes E and F contain the list of all the documents we coded during our literature review; Appendix E includes journal articles we reviewed, and Appendix F includes codes of conduct. Lastly, the References section includes all documents we cite in this report that were not part of our literature review and do not appear in Appendixes E or F.

How This Report Could Be Used

This report could be valuable to researchers, ethicists, scientific societies, sponsors of research, and government regulators seeking to create new codes of ethics for emerging disciplines or struggling to decipher how ethics should apply to a new situation or discipline. For the experienced researchers, this report provides perspective outside of their own fields across the breadth of disciplines and ethical challenges they have yet to encounter. As more researchers collaborate across fields, sometimes helping to shape new ones, this report could aid those who find themselves working on research not contemplated by their formal training and past experiences. For instance, it is common today for geneticists to work with big data and information science; for computer scientists to conduct sociological studies on social media; and for those who may lack formal training to engage in research that historically required considerable training. For this last group, the citizen scientist or amateur researcher, this report provides guideposts to understanding ethical norms, behaviors, and practices.

The Difference Between Ethics and Law

Laws are geographically based *and* biased by local cultural norms. Each country, state, and locality can pass its own laws legalizing or banning any behavior. Ethics, on the other hand, reflect the values of a collective—a population, at their most general, or a professional society or other group in specific instances. Ethics may or may not agree with local laws. This sometimes-difficult relationship is described succinctly in the preamble to the American Medical Association's Code of Ethics:

> The relationship between ethics and law is complex. Ethical values and legal principles are usually closely related, but ethical responsibilities usually exceed legal duties. Conduct that is legally permissible may be ethically unacceptable. Conversely, the fact that a physician who has been charged with allegedly illegal conduct has been acquitted or exonerated in criminal or civil proceedings does not necessarily mean that the physician acted ethically.

> In some cases, the law mandates conduct that is ethically unacceptable. When physicians believe a law violates ethical values or is unjust they should work to change in law. In exceptional circumstances of unjust laws, ethical responsibilities should supersede legal duties.[1]

Another perspective comes from consideration of why codes are produced by professional societies: The organization of a profession is typically accompanied by a code, which establishes a convention among people in a field that helps to define who they are as professionals:

> A code of ethics . . . prescribe[s] how professionals are to pursue their common ideal so that each may do the best she can at minimal cost to herself and those she cares about (including the public, if looking after the public is part of what she cares about). The code is to protect each professional from certain pressures (for example, the pressure to cut corners to save money) by making it reasonably likely (and more likely than otherwise) that most other members of the profession will not take advantage of her good conduct. A code protects members of a profession

---

[1] American Medical Association, "Code of Medical Ethics: Preface and Preamble," 2016.

from certain consequences of competition. A code is a solution to a coordination problem.[2]

Thus, a code provides guidance beyond what might come from personal conscience alone.

This project focused on ethics—not laws or regulations. We examined ethics across international borders to understand which ethical principles or elements are shared—across disciplines, countries, and societies—and to identify nuances within ethical elements. Throughout our research, we found examples of professional societies that hold their members to ethical standards (sometimes called "soft law") independent of legal standards. Additionally, we found examples of professional societies that required researchers to behave ethically by abiding by national law.

---

[2] M. Davis, "Thinking Like an Engineer: The Place of a Code of Ethics in the Practice of a Profession, *Philosophy and Public Affairs*, Vol. 20, No. 2, 1991, pp. 150–167.

CHAPTER TWO

Ethical Principles for Scientific Research

Our team collected the 200 most-cited peer-reviewed articles on ethics in scientific research from literature over time plus codes of ethics from across scientific disciplines. We then analyzed these documents for commonalities and differences, and we present our results in this chapter. Where we found gaps in the literature, we conducted additional research and interviews with experts in relevant fields.[1] As a result of our analysis, we identified ten ethical principles that cross scientific and technical disciplines. We present this list in Table 2.1 with our definitions for each term, and throughout the remainder of the chapter, we explain what the literature said on each topic in greater detail.

Each ethical principle applies to one or more of the following categories:

- ethical scientific inquiry
- ethical conduct and behaviors of researchers
- ethical treatment of research participants.

Table 2.2 describes the categories and maps the ethical principles to them. One principle, nondiscrimination, applies in two categories: behaviors of researchers and treatment of participants. Although the principles are discussed as discrete sets of concerns, they interconnect, if not overlap.

Duty to Society

Definition: *Researchers and research must contribute to the well-being of society.*

Duty to society is a well-documented element of ethics across our literature review, and yet it differs slightly between disciplines and countries. International differences will be discussed in Chapter Three. The primary premise of duty to society is

---

[1] Appendix A includes a detailed description of our methodology, including how we chose which articles to review and which experts to interview and how we conducted our interviews and our analysis.

Table 2.1
Ethical Principles for Scientific Research

Ethical Principle	Definition
Duty to society	Researchers and research must contribute to the well-being of society.
Beneficence	Researchers should have the welfare of the research participant in mind as a goal and strive for the benefits of the research to outweigh the risks.
Conflict of interest	Researchers should minimize financial and other influences on their research and on research participants that could bias research results. Conflict of interest is more frequently directed at the researcher, but it may also involve the research participants if they are provided with a financial or nonfinancial incentive to participate.
Informed consent	All research participants must voluntarily agree to participate in research, without pressure from financial gain or other coercion, and their agreement must include an understanding of the research and its risks. When participants are unable to consent or when vulnerable groups are involved in research, specific actions must be taken by researchers and their institutions to protect the participants.
Integrity	Researchers should demonstrate honesty and truthfulness. They should not fabricate data, falsify results, or omit relevant data. They should report findings fully, minimize or eliminate bias in their methods, and disclose underlying assumptions.
Nondiscrimination	Researchers should minimize attempts to reduce the benefits of research on specific groups and to deny benefits from other groups.
Nonexploitation	Researchers should not exploit or take unfair advantage of research participants.
Privacy and confidentiality	**Privacy:** Research participants have the right to control access to their personal information and to their bodies in the collection of biological specimens. Participants may control how others see, touch, or obtain their information. **Confidentiality:** Researchers will protect the private information provided by participants from release. Confidentiality is an extension of the concept of privacy; it refers to the participant's understanding of, and agreement to, the ways identifiable information will be stored and shared.
Professional competence	Researchers should engage only in work that they are qualified to perform, while also participating in training and betterment programs with the intent of improving their skill sets. This concept includes how researchers choose research methods, statistical methods, and sample sizes that are appropriate and would not cause misleading results.
Professional discipline	Researchers should engage in ethical research and help other researchers engage in ethical research by promulgating ethical behaviors through practice, publishing and communicating, mentoring and teaching, and other activities.

NOTE: *Research participant* refers to someone with an active role participating in research, whereas *research subject* could include someone whose data are used but who does not consent to participate.

that research must not be undertaken if it produces no benefit to society.[2] Such benefit is judged by the researchers, their institution, and their sponsors, rather than by society as a whole or by historians in future decades, leading to lapses between what research-

---

[2]   Interview 11. The National Society of Professional Engineers, although focused on professional practice rather than research, has a similar concept in the *Paramountcy Principle*—holding paramount the safety, health, and welfare of the public. See National Society of Professional Engineers, *Code of Ethics*, Alexandria, Va., 2018.

Table 2.2
Categories of Ethical Principles

Category	Description of Category	Ethical Principle
Ethical scientific inquiry	The research inquiry itself must benefit society.	• Duty to society
Ethical conduct and behaviors of researchers	Researchers should conduct themselves in certain manners, and they are responsible for their knowledge and awareness of ethics and appropriate research methods.	• Conflict of interest • Integrity • Nondiscrimination • Professional competence • Professional discipline
Ethical treatment of research participants	Research participants should be treated according to certain guidelines and treated humanely, and the environmental or secondary effects of the research should be considered.	• Informed consent • Beneficence • Nondiscrimination • Nonexploitation • Privacy and confidentiality

ers and the research community believe is a benefit to society and what other members of society might believe.

Some unethical activities conducted in the name of medical research involved the inhumane treatment of research participants without a broader benefit to society or with benefits that could not have been foreseen at the time. In some cases, duty to society comes in conflict with beneficence, as when society may benefit from research that may knowingly and intentionally harm research participants. Historical examples provide cases where society has benefited from research that was inhumane to its participants, and scientists still grapple today with whether it is ethical to use the results of such research. One researcher calculated that by 2010, "the data from Nazi experiments have been used and/or cited on over fifty occasions," particularly "data from hypothermia experiments."[3] In modern ethics, both beneficence and duty to society are simultaneously required: Research must benefit or aim to do no harm to both the research subjects and society.[4] There is no universal equilibrium, since some cultures place more emphasis on the well-being of a community over that of the individual.[5] Involving members of any community can help in designing research that achieves an appropriate balance.

---

[3] R. Halpin, "Can Unethically Produced Data Be Used Ethically?" *Medicine and Law*, Vol. 29, 2010, pp. 373–387.

[4] E. J. Emanuel, D. Wendler, and C. Grady, "What Makes Clinical Research Ethical?" *Journal of the American Medical Association*, Vol. 283, No. 20, 2000, pp. 2701–2711.

[5] Interviews 1 and 11.

In medical disciplines, the literature states that the primary obligation of researchers should be to their participants, not to the objectives of their studies.[6] This principle was documented in the first version of the Declaration of Helsinki in 1964 (see Chapter Three), which said, "[c]linical research cannot legitimately be carried out unless the importance of the objective is in proportion to the inherent risk to the subject."[7]

Violations of this ethical principle can occur when research is conducted in countries where regulations are less stringently observed (see discussion of "ethics dumping" in Chapter Three). Emanuel et al. equates ethical multinational research with avoidance of exploitation, the risk of which is greater in developing countries. Key to avoiding exploitation is a collaborative partnership in such contexts to reinforce other ethical principles.[8]

Nonmedical Guidelines

Each discipline we examined had slightly different interpretations or applications of duty to society. None of these interpretations conflicted with the others, but each related to the specific needs of researchers in that specific field. In genomics, members of society are responsible for determining the appropriate and inappropriate use of genetic research.[9] Researchers must consider the perspectives of diverse communities across society to understand their ethical boundaries, values, and concerns on how this discipline affects society.[10] Overall, "respect for the dignity and well-being of persons takes precedence over expected benefits to knowledge."[11]

According to the International Society of Ethnobiology,

> persons and organizations undertaking research activities shall do so throughout in good faith, acting in accordance with, and with due respect for, the cultural norms and dignity of all potentially affected communities, and with a commitment that collecting specimens and information, whether of a zoological, botanical, mineral or cultural nature, and compiling data or publishing information thereon, means

---

[6] M. Angell, "The Ethics of Clinical Research in the Third World," *New England Journal of Medicine*, Vol. 337, No. 12, 1997, pp. 847–849; B. Freedman, "Equipoise and the Ethics of Clinical Research," *New England Journal of Medicine*, Vol. 317, No. 3, 1987, pp. 141–145; Emanuel, Wendler, and Grady, 2000.

[7] World Medical Association, "DECLARATION OF HELSINKI: Recommendations Guiding Doctors in Clinical Research," adopted by the 18th World Medical Assembly, Helsinki, Finland, June 1964.

[8] E. J. Emanuel, D. Wendler, J. Killen, and C. Grady, "What Makes Clinical Research in Developing Countries Ethical? The Benchmarks of Ethical Research," *Journal of Infectious Diseases*, Vol. 189, No. 5, 2004, pp. 930–937.

[9] F. S. Collins, E. D. Green, A. E. Guttmacher, and M. S. Guyer, "A Vision for the Future of Genomics Research," *Nature*, Vol. 422, No. 6934, 2003, pp. 835–847.

[10] Collins et al., 2003; M. Minkler, "Community-Based Research Partnerships: Challenges and Opportunities," *Journal of Urban Health*, Vol. 82 (SUPPL. 2), 2005, pp. ii3–ii12.

[11] M. Guillemin and L. Gillam, "Ethics, Reflexivity, and 'Ethically Important Moments' in Research," *Qualitative Inquiry*, Vol. 10, No. 2, 2004, pp. 261–280.

doing so only in the holistic context, respectful of norms and belief systems of the relevant communities.[12]

In engineering, where apart from bioengineering and some aspects of computer systems, research rarely includes research participants, duty to society assigns engineers responsibility for the safety of the public.[13] In the ethical code of the Association for Computing Machinery (ACM), a professional society for computer scientists, the first principle states, "[c]ontribute to society and to human well-being, acknowledging that all people are stakeholders in computing."[14] And in ecology, researchers should strive to understand the complex relationship between biodiversity ecosystem functioning and management to minimize current losses of species and responsibly manage Earth's ecosystems.[15] Here, society is not even limited to humans, but rather includes all of Earth's ecosystems.

We found that the literature involving human participants prioritizes participants' well-being over the potential knowledge gained or the benefits expected from research; second, research design should consider diverse perspectives for how a project may affect a population. One issue that researchers may grapple with is defining the process of protecting public welfare or safety. Because this standard is subjective, viewpoints may differ on how to comply with such a mandate. In disciplines that are not human-centric, we found that researchers are urged to highlight the need to minimize the harm to our environment.

Some codes of conduct go even further to define specific activities they deem unethical to society, and in these instances these codes ban researchers from participating in such activities. Notable examples our team found include the following:

- The American Society for Microbiology (ASM) instructs members "to discourage any use of microbiology contrary to the welfare of humankind, including the use of microbes as biological weapons. Bioterrorism violates the fundamental principles upon which the Society was founded and is abhorrent to the ASM and its members."[16]
- The International Sociological Association (ISA) warns members to be vigilant of sponsors who wish to use research for "political aims." The association says soci-

---

[12] International Society of Ethnobiology, "ISE Code of Ethics," 2008.

[13] L. J. Shuman, S. M. Besterfield-Sacre, and J. McGourty, "The ABET 'Professional Skills': Can They Be Taught? Can They Be Assessed?" *Journal of Engineering Education*, Vol. 94, No. 1, 2005, pp. 41–55.

[14] ACM, "ACM Code of Ethics and Professional Conduct," 2018.

[15] D. U. Hooper, F. S. Chapin III, J. J. Ewel, A. Hector, P. Inchausti, S. Lavorel, J. H. Lawton, D. M. Lodge, M. Loreau, S. Naeem, B. Schmid, H. Setälä, A. J. Symstad, J. Vandermeer, and D. A. Wardle, "Effects of Biodiversity on Ecosystem Functioning: A Consensus of Current Knowledge," *Ecological Monographs*, Vol. 75, No. 1, 2005, pp. 3–35.

[16] ASM, "Code of Ethics," 2005.

ologists "should also refrain from cooperating in the fulfillment of undemocratic aims or discriminatory goals."[17]

- The International Society for Environmental Epidemiology advocates for research that "place[s] the health of exposed or at-risk populations ahead of concern for the reputation and financial well-being of any institution or organization."[18]

All professional societies that we studied instruct their members to fundamentally serve the public with the fruits of their research and practice. Some societies provide guidance on unique issues when attempting to comply with that mission. The vast majority of statements about duty to society focus on protecting public welfare. The ASM, ISA, and the International Society for Environmental Epidemiology are unique examples from our research in how they specify the threats that their discipline could pose to the public by making explicit what constitutes *unethical* research and behavior.

In modern research, duty to society continues to exert ethical dilemmas in emerging research disciplines, such as information science research conducted on society-wide data sets. In some cases, it is possible to ask whether the benefits to society for such research will ever be realized. See Box 2.1 for a discussion of the Uppsala Code, which calls on researchers to ponder the prospective societal impacts as individuals and eschew research that could support war or oppression.

Beneficence

Definition: *Researchers should have the welfare of the research participant in mind as a goal and strive for the benefits of the research to outweigh the risks.*

Beneficence is a core tenet of any research that involves human participants, and, as such, it could be called a pillar of medical research. Simply put, beneficence requires that research be designed to maximize the benefits to research participants while minimizing the harm to them. According to the literature, the benefits of the research may not be artificially inflated by researchers to disguise the harms nor to offset the severity of the harms.[19] In other words, any financial or nonfinancial benefits offered to research participants—including payment for participation, free medical tests, free medical exams, free vaccinations, and so on—cannot be considered in an assessment of beneficence. Accordingly, monitoring boards, including institutional review boards

---

[17] International Sociological Association, "Code of Ethics," 2001.

[18] International Society for Environmental Epidemiology, "Ethics Guidelines for Environmental Epidemiologists," 2012.

[19] G. B. Drummond, "Reporting Ethical Matters in the *Journal of Physiology*: Standards and Advice," *Journal of Physiology*, Vol. 587, No. 4, 2009, pp. 713–719; Emanuel, Wendler, and Grady, 2000; T. S. Behrend, D. J. Sharek, A. W. Meade, and E. N. Wiebe, "The Viability of Crowdsourcing for Survey Research," *Behavior Research Methods*, Vol. 43, No. 3, 2011, pp. 800–813.

Box 2.1
The Uppsala Code and the Pugwash Tradition

The Uppsala (Sweden) Code of Ethics for Scientists was developed in the early 1980s to address the potential for scientists to influence the balance between war and peace (and to protect the environment), since research can either ameliorate or aggravate problems in society. Notwithstanding these broad concerns, the Uppsala Code focuses on the responsibilities of individual scientists: "We consider the ethical dilemmas that the code addresses to be personal ones; they are matters of conscience." This code invokes a "duty to inform": "When a scientist finds his/her own work unethical he/she should interrupt it."

The Uppsala Code drew from Pugwash discussions, which involve scientists exploring how they can contribute to evidence-based policymaking intended to combat the threat of weapons of mass destruction. Whereas much of the discussion so far has addressed protections for individuals and definable groups, the Uppsala Code has a broader societal focus. According to the people spearheading its development, "A code should give some details about the responsibility of the scientist and some advice on how to act when an ethical dilemma arises." They observed that codes of ethics associated with research typically are written in a general way. That circumstance opens the door to both interpretation and ambiguity, both of which can undercut [implementation]. The developers of the Uppsala Code also put a spotlight on the responsibility of scientists to consider how their work might be used.

SOURCES: Bengt Gustafsson, Lars Rydén, Gunnar Tybell, and Peter Wallensteen, "Focus on: The Uppsala Code of Ethics for Scientists," *Journal of Peace Research*, Vol. 21, No. 4, 1984, pp. 311–316; Pugwash, "Pugwash Conferences on Science and World Affairs," homepage, 2018.

(IRBs) that evaluate the continuation of research on human participants, have become an important element of ensuring beneficence.[20]

In medicine, adherence to the principle of beneficence reconciles the tensions between the responsibility to provide a quality of care and the need for research to test new treatments by requiring researchers to hold the welfare of the research participant to the highest standards. Thus, for example, researchers must consider whether using a placebo or untreated control group is ethical when effective treatments exist. The ethics of using placebos during medical research are still under debate;[21] placebos are generally accepted when they present low risk and are essential to a methodology. One notorious example in which researchers did not take the standard of care into account was the "Tuskegee Study of Untreated Syphilis in the Negro Male," whose research participants were allowed to suffer the effects of untreated syphilis despite an effec-

---

[20] An *IRB* is a group that has formal, designated authority to review and monitor research involving human subjects. An IRB has the authority to approve, require modifications in, or disapprove research. This group review serves an important role in the protection of the rights and welfare of human research subjects. The purpose of IRB review is to assure, both in advance and by periodic review, that appropriate steps are taken to protect the rights and welfare of humans participating as subjects in the research. To accomplish this purpose, IRBs use a group process to review research protocols and related materials (e.g., informed consent documents and investigator brochures) to ensure protection of the rights and welfare of human subjects of research (U.S. Food and Drug Administration, "Institutional Review Boards Frequently Asked Questions: Information Sheet," fact sheet, July 12, 2018).

[21] A. Skierka and K. Michels, "Ethical Principles and Placebo-Controlled Trials: Interpretation and Implementation of the Declaration of Helsinki's Placebo Paragraph in Medical Research," BMC Medical Ethics, 2018; J. Millum and C. Grady, "The Ethics of Placebo-Controlled Trials: Methodological Justifications," *Contemporary Clinical Trials*, Vol. 36, No. 2, 2013, pp. 510–514.

tive treatment existing during the study.[22] The Tuskegee study has become the case example of research that lacked beneficence. Research literature discusses the need to provide high-quality care to research participants who might not otherwise receive it due to geographic location, illiteracy, poverty, or other factors.

In clinical contexts, as opposed to research contexts, a physician is expected to be guided by both beneficence and the complementary concept of avoiding harm (nonmaleficence). Because research involves more uncertainty than clinical care—reducing uncertainty is a goal of research—it is understood that there is a risk of harm (see the discussion of informed consent later in this chapter), which should be outweighed by the potential for benefit.

The Declaration of Helsinki, the history of which is discussed in Chapter Three, states that "[i]t is the duty of physicians who are involved in medical research to protect the life, health, dignity, integrity, right to self-determination, privacy, and confidentiality of personal information of research subjects."[23] This usage of integrity as a component of beneficence "includes respect for the autonomy of individuals, achieved mainly by the mechanism of informed consent; respect for privacy, achieved at least partly by rules relating to confidentiality and secure storage of data; and respect for the dignity of persons."[24] (Integrity as an attribute of researcher behavior is discussed later in this chapter.)

Beneficence is also closely linked to informed consent (another topic discussed later in this chapter). "The subjects should be volunteers," says the Declaration of Helsinki, which goes on to state, "[w]hile the primary purpose of medical research is to generate new knowledge, this goal can never take precedence over the rights and interests of individual research subjects."[25] A research participant should not be asked to consent to a study that lacks sufficient benefits, and informed consent does not replace the need for beneficence.[26] Incentives to research participants, such as financial payments, free vaccines, or medications, should not be used to tip the scale of beneficence, making the benefits of the study appear to outweigh the harms to the research participants. Nor should these benefits be used to coerce informed consent.

Research studies have been conducted in which research participants have experienced psychological stress and other negative effects that can be unexpected by

---

22   Centers for Disease Control and Prevention, "U.S. Public Health Service Syphilis Study at Tuskegee," 2015.

23   WMA, "Declaration of Helsinki: Ethical Principles for Medical Research Involving Human Subjects," *Journal of the American Medical Association*, Vol. 310, No. 20, 2013.

24   Guillemin, 2004.

25   WMA, 2013.

26   K. J. Rothman and K. B. Michels, "The Continuing Unethical Use of Placebo Controls," *New England Journal of Medicine*, Vol. 331, No. 6, 1994, pp. 394–398.

researchers.[27] It is the responsibility of researchers to consider possible harm that may come to a participant and to respond when new unexpected harms occur.[28]

Research that uses human subjects' data—without experimenting on research participants themselves—is similarly required to adhere to beneficence. Even if data are "de-identified," the literature says they "must be justified to the IRB as having some expected benefits. . . . One cannot perform data analysis for frivolous or nefarious purposes." This applies to vast web-based data sets that may be publicly available (including social media data).[29] Challenges associated with big data are addressed in Chapter Five.

Conflicts of interest, a major area of concern in their own right discussed later in this chapter, can affect beneficence. As described by the American Psychological Association (APA):

> Psychologists strive to benefit those with whom they work and take care to do no harm. In their professional actions, psychologists seek to safeguard the welfare and rights of those with whom they interact professionally and other affected persons, and the welfare of animal subjects of research. When conflicts occur among psychologists' obligations or concerns, they attempt to resolve these conflicts in a responsible fashion that avoids or minimizes harm. Because psychologists' scientific and professional judgments and actions may affect the lives of others, they are alert to and guard against personal, financial, social, organizational, or political factors that might lead to misuse of their influence. Psychologists strive to be aware of the possible effect of their own physical and mental health on their ability to help those with whom they work. [30]

APA continues, "[p]sychologists take reasonable steps to avoid harming their clients/patients, students, supervisees, research participants, organizational clients,

---

[27] B. DiCicco-Bloom and B. F. Crabtree, "The Qualitative Research Interview," *Medical Education*, Vol. 40, No. 4, 2006, pp. 314–321.

[28] For research on animals, literature on beneficence requires minimizing pain and using anesthesia and "pain blocking agents." Animals should be housed and fed in humane conditions, and different levels of protections are applied to cats, dogs, primates, and horses versus invertebrates and "lower levels" of species. In some countries (such as the United Kingdom), these rules are written in law, and violations are punishable with prison sentences (Drummond, 2009). Guidelines for use of wild mammal species are updated from an American Society of Mammalogists 2007 publication (R. S. Sikes and W. L. Gannon, "Guidelines of the American Society of Mammalogists for the Use of Wild Mammals in Research," *Journal of Mammalogy*, Vol. 92, No. 1, 2011, pp. 235–253). Of course, reasonable members of society can debate whether research conducted on animals is ever humane, yet we found no literature that required the researcher to adjudicate this ethical debate. Instead, the literature requires researchers to treat animals humanely and with beneficence and often stipulates that animals should be used only when necessary, though such necessity is rarely, if ever, defined.

[29] K. J. Cios and G. W. Moore, "Uniqueness of Medical Data Mining," *Artificial Intelligence in Medicine*, Vol. 26, No. 1–2, 2002, pp. 1–24.

[30] APA, "American Psychological Association's Ethical Principles of Psychologists and Code of Conduct," 2017.

and others with whom they work, and to minimize harm where it is foreseeable and unavoidable."

In the ACM code of ethics for computer scientists, the first principle states, "[c]ontribute to society and to human well-being." Its second principle states simply, "[a]void harm," which is defined thus:

> In this document, "harm" means negative consequences, especially when those consequences are significant and unjust. Examples of harm include unjustified physical or mental injury, unjustified destruction or disclosure of information, and unjustified damage to property, reputation, and the environment. This list is not exhaustive.

> Well-intended actions, including those that accomplish assigned duties, may lead to harm. When that harm is unintended, those responsible are obliged to undo or mitigate the harm as much as possible. Avoiding harm begins with careful consideration of potential impacts on all those affected by decisions. When harm is an intentional part of the system, those responsible are obligated to ensure that the harm is ethically justified. In either case, ensure that all harm is minimized.[31]

The American Statistical Association says researchers should "[Strive] to avoid the use of excessive or inadequate numbers of research subjects—and excessive risk to research subjects (in terms of health, welfare, privacy, and ownership of their own data)—by making informed recommendations for study size."[32]

Our analysis suggested that beneficence may be in the eye of the beholder for both research participants and researchers. Without clear guidelines on how to achieve it, beneficence is described in cost-benefit terms comparing human participants' gains against the risk of harm that they might incur, and the research participant makes this assessment when asked to give his or her informed consent. The various codes use phrases like "safeguard his or her integrity" (medicine), "recognizes the autonomy of individuals" (general sciences), "minimize harm where it is foreseeable and unavoidable" (psychology), and "consider the potential impact" (social science). These are not clear distinctions between research that is good (beneficent) and bad (maleficent). Rather, these statements call on researchers to use their best judgment and honor for the sake of their participants. Similarly, in the engineering field, the challenge to work under cost and schedule pressures can increase risks. Engineers have a responsibility to ask when the risk has increased such that it is no longer acceptable.[33]

An interesting exception, where beneficence is defined in black-and-white terms rather than being left to the interpretation of the researcher, can be found in the APA's

---

[31] ACM, 2018.

[32] American Statistical Association, "Ethical Guidelines for Statistical Practice," 2018.

[33] Shuman, Besterfield-Sacre, and McGourty, 2005.

code of conduct, which specifically calls on psychologists not to "participate in, facilitate, assist, or otherwise engage in torture." To prevent any misunderstanding, the code defines torture as "any act by which severe pain or suffering, whether physical or mental, is intentionally inflicted on a person, or in any other cruel, inhuman, or degrading behavior that violates 3.04(a).3." This level of specificity is rare among codes of conduct and is an issue the APA has publicized extensively.[34]

We found widespread agreement across the literature on what beneficence means, though certain disciplines add additional insights and nuance as to how it applies in their respective fields. Disciplines that do not engage routinely with research participants—such as computer science/information science and mathematics/statistics—lack comprehensive literature on this topic, despite the potential for their research to affect persons whose data are used, analyzed, and reported on. This is one reason why we make a distinction in this report between research participants (who have an active participatory role in the research) and research subjects (who do not consent to participate but may be affected by research or whose data may be used). Technology fields, including computer science, are more recently coming to terms with their role in affecting human *subjects* and their ethical responsibility to beneficence. Given that we focused our research on highly cited articles, we would anticipate that newer literature has or could address this perceived gap. The recent emergence of codes of ethics and related scholarship for researchers working on AI illustrates a corresponding evolution in thinking.

Conflict of Interest

Definition: *Researchers should minimize financial and other influences on their research and on research participants that could bias research results. Conflict of interest is more frequently directed at the researcher, but it may also involve the research participants if they are provided with a financial or nonfinancial incentive to participate.*

The National Academies of Sciences, Engineering, and Medicine describe conflicting interests thus: "in some cases the prospect of financial gain could affect the design of an investigation, the interpretation of data, or the presentation of results. Indeed, even the appearance of a financial conflict of interest can seriously harm a researcher's reputation as well as public perceptions of science."[35] Across our research, we found literature that sets guidelines for how to identify conflicts while acknowledging that they may arise in many forms.

---

[34] See APA, *Timeline of APA Policies and Actions Related to Detainee Welfare and Professional Ethics in the Context of Interrogation and National Security*, Washington, D.C., 2019b; APA, *Position on Ethics and Interrogation*, Washington, D.C., 2019a.

[35] National Academy of Sciences, National Academy of Engineering, and Institute of Medicine of the National Academies, *On Being a Scientist: A Guide to Responsible Conduct in Research*, 3rd Edition, 2009.

The literature documents how conflicts of interest can be financial or nonfinancial (including the provision of equipment, services, speaking and publishing opportunities, professional opportunities, or any other personal gain to the researcher). Codes of conduct that discuss conflicts of interest place responsibility on researchers to prevent and/or disclose any such relationships. Many journals require such disclosure of support for their research from authors prior to accepting articles for publication.

For research participants who are paid for their participation, the payment itself, as well as any nonmonetary benefits of participating, can create a conflict in preventing the participant from accurately weighing the risks and benefits of the research. In this sense, a financial or nonfinancial benefit for participating (including free medical exams, free medical tests, free vaccinations, and so on), can influence whether a research participant provides an uncoerced consent to participate. In this sense, any financial or nonfinancial benefits to research participants should be evaluated by IRBs or other oversight boards, as both the research participant and researcher may be unable to assess the potential coercive effect of the benefit without bias.

Undisclosed conflicts of interest could cast doubt on the validity of the data, the analysis, the selection of research participants, the public's trust in research, and other factors. The literature addresses conflicts relating to either the funder or the research participant. Conflicts associated with the nature of funders are widely discussed, especially throughout medical research literature, since a significant amount of that research is funded by pharmaceutical companies, medical device companies, and other for-profit entities that may benefit from findings.

Conflicts of interest can also affect what, when, and how the results of research are published and therefore benefit the larger research community. Our review found reports "that studies sponsored by a pharmaceutical company were less likely to be published, whatever the results" and "selectivity in the submission for publication of drug company sponsored studies, according to the direction of the result."[36] In neurosurgery specifically, researchers found, "Industry funding was associated with a much greater chance of positive findings in [randomized controlled trials] published in neurosurgical journals. Further efforts are needed to define the relationship between the authors and financial sponsors of neurosurgical research and explore the reasons for this finding."[37] In an article titled, "Scope and Impact of Financial Conflicts of Interest in Biomedical Research," researchers said,

---

[36] P. J. Easterbrook, R. Gopalan., J. A. Berlin, and D. R. Matthews, "Publication Bias in Clinical Research," *The Lancet*, Vol. 337, No. 8746, 1991, pp. 867–872.

[37] N. R. Khan, H. Saad, C. S. Oravec, N. Rossi, V. Nguyen, G. T. Venable, J. C. Lillard, P. Patel, D. R. Taylor, B. N. Vaughn, D. Kondziolka, F. G. Barker, L. M. Michael, and P. Klimo, "A Review of Industry Funding in Randomized Controlled Trials Published in the Neurosurgical Literature: The Elephant in the Room, *Neurosurgery*, Vol. 83, No. 5, 2018, pp. 890–897.

Consistent evidence also demonstrated that industry ties are associated with both publication delays and data withholding. These restrictions, often contractual in nature, serve to compound bias in biomedical research. Anecdotal reports suggest that industry may alter, obstruct, or even stop publication of negative studies. Such restrictions seem counterproductive to the arguments in favor of academic industry collaboration, namely encouraging knowledge and technology transfer. Evidence shows, however, that industry sponsorship alone is not associated with data withholding. Rather, such behavior appears to arise when investigators are involved in the process of bringing their research results to market.[38]

Minor gift-giving is even less monitored and governed than industry sponsorship of research. Even minor gifts—well below any reporting threshold—can create loyalty and bias for the recipient to the giver or lead to an expectation of reciprocity, and some argue that gifts of any size should be banned.[39] In lieu of tangible gifts, some medical companies and industry groups offer speaking or publication opportunities to researchers, which present nonfinancial conflicts that can aggravate the pressure to be recognized and the bias to publish only research with positive results. This is especially risky when the gift is nonfinancial and not easily recognized as a gift, such as a professional opportunity, speaking opportunity, or other nonfinancial benefit.[40]

Conflicts of interest between researcher and research participants are less widely discussed within the ethical principle of conflict of interest yet present their own challenges. In the social sciences, two articles in our data set addressed conflicts that arise when the researcher has or develops a personal relationship with the subject. The potential results of personal relationships could affect whether researchers make the best judgments for the research rather than for their friend the research participant.[41] In the medical field, the American Medical Association warns its physicians to be careful of conflict when a patient becomes a research participant in the physician's own clinical trial: the physician will have conflicting loyalty to the patient and the research.[42] When

---

[38] J. E. Bekelman, Y. Li, and C. P. Gross, "Scope and Impact of Financial Conflicts of Interest in Biomedical Research: A Systematic Review," *Journal of the American Medical Association*, Vol. 289, No. 4, 2003, pp. 454–465.

[39] T. A. Brennan, D. J. Rothman, L. Blank, D. Blumenthal, S. C. Chimonas, J. J. Cohen, J. Goldman, J. P. Kassirer, H. Kimball, J. Naughton, and N. Smelser, "Health Industry Practices That Create Conflicts of Interest: A Policy Proposal for Academic Medical Centers," *Journal of the American Medical Association*, Vol. 295, No. 4, 2006, pp. 429–433.

[40] A. W. Chan, J. M. Tetzlaff, P. C. Gøtzsche, D. G. Altman, H. Mann, J. A. Berlin, K. Dickersin, A. Hróbjartsson, K. F. Schulz, W. R. Parulekar, K. Krleza-Jeric, A. Laupacis, and D. Moher, "SPIRIT 2013 Explanation and Elaboration: Guidance for Protocols of Clinical Trials," *British Medical Journal* (Clinical Research Edition), Vol. 346, No. 2, 2013.

[41] See C. Ellis, "Telling Secrets, Revealing Lives: Relational Ethics in Research with Intimate Others," *Qualitative Inquiry*, Vol. 13, No. 1, 2007, pp. 3–29; Behrend, 2011.

[42] American Medical Association, 2016, chapter 7; American Medical Association, "Principles of Medical Ethics," 2018.

the researcher has a vested interest in the research participant, such as via the doctor-patient relationship, the researcher may be divided in loyalties between wanting the patient to live a good quality of life and wanting to conduct unbiased research.

Across the literature, we found widespread agreement that conflicts of interest should be addressed in a timely manner, managed by the researcher(s) and institution(s), and disclosed to the public and the research participants.[43] And yet disclosure to research participants and the public may be insufficient: Can the researcher know the extent to which the conflict affected his or her results? How would a research participant appreciate the significance of the conflict? And how would other experts in the field be able to judge the validity of the research without reproducing the results? Research participants may have a strong desire for continuing in the research treatment and be unable to judge objectively the importance of the conflict. Meanwhile, outside readers of published research results may be unable to adequately judge the effect the conflict had on the research.

Some professional societies mitigate these risks by providing specific guidance, recommendations, or disclosure requirements to their researchers.

- The Society of Toxicology provides disclosure forms, documents, and definitions for researchers.[44]
- The Ethical Standards in Sport and Exercise Science Research requires authors to "include details of any incentives for participants and provisions for treating and/ or compensating participants who are harmed as a consequence of participation in the research study."[45]
- The American Association of Physicists in Medicine informs its members that NIH has established a reporting requirement of financial gains of $10,000 or more and addresses nonfinancial gains, such as prestige.[46]
- In the business sector, we found that the Academy of Management tells its members to avoid both actual conflicts of interest and the *appearances* of conflicts.[47]

---

[43] National Academy of Sciences, National Academy of Engineering, and Institute of Medicine of the National Academies, 2009.

[44] Links to these resources are at the bottom of Society of Toxicology, "Code of Ethics and Conflict of Interest," webpage, last revised 2012.

[45] D. J. Harriss and G. Atkinson, "Ethical Standards in Sport and Exercise Science Research: 2014 Update," *International Journal of Sports Medicine*, Vol. 34, No. 12, 2013, pp. 1025–1028.

[46] American Association of Physicists in Medicine, *Code of Ethics for the American Association of Physicists in Medicine: Report of Task Group 109*, 2009.

[47] Academy of Management, "AOM Code of Ethics," December 6, 2017.

- An international code for sociologists warns its members of sponsors "interested in funding sociological research for the sake of their own political aims. Whether they share such aims, sociologists should not become subordinate to them."[48]
- In anthropology, "Anthropologists have an obligation to distinguish the different kinds of interdependencies and collaborations their work involves, and to consider the real and potential ethical dimensions of these diverse and sometimes contradictory relationships, which may be different in character and may change over time." The code specifically calls out "obligations to vulnerable populations" and declares that "Anthropologists remain individually responsible for making ethical decisions."[49]
- The International Society for Environmental Epidemiology holds that members should "not accept funding from sponsors, accept contractual obligations, or engage in research that is contingent upon reaching particular conclusions from a proposed environmental epidemiology inquiry."[50]

Across disciplines, we found varied discussions about what types of activities, relationships, or behaviors constitute or indicate a possible conflict of interest. At the same time, we found less discussion about the difference between disclosing conflicts and managing conflicts. Most articles focus on identifying the conflict—how to know it when it occurs—and places the responsibility on researchers to disclose or mitigate them. But these same literature sources lack sufficient and useful strategies for researchers trying to manage conflicts in a world awash with industry funding and interconnected relationships.

Informed Consent

Definition: *All research participants must voluntarily agree to participate in research, without pressure from financial gain or other coercion, and their agreement must include an understanding of the research and its risks. When participants are unable to consent or when vulnerable groups are involved in research, specific actions must be taken by researchers and their institutions to protect the participants.*

Informed consent may be one of the best-defined ethical elements across our research. Every discipline we examined that uses human research participants agrees on the need for informed consent, and we found few variations in how it should be applied. We found widespread agreement in the literature that informed consent

---

48 International Sociological Association, 2001.

49 American Anthropological Association, "Principles of Professional Responsibility," November 1, 2012.

50 International Society for Environmental Epidemiology, 2012.

- must use language the research participant understands and comprehends to explain the research, its risks to the participant, and its benefits to the participant
- must be given freely by the research participant
- may be revoked by the research participant at any time
- may only be asked of and given by adults who are capable of making an informed consent, and when research participant are neither adults nor capable of making an informed consent, review boards should provide oversight to protect the rights of research participants, which could involve engaging surrogates or proxies (who raise their own issues).[51]

Debate on this topic begins to arise in two areas, the first of which regards whether pregnant women should be considered a vulnerable group. U.S. federal regulations define three groups of vulnerable populations in research: pregnant women, human fetuses, and neonates; children; and prisoners.[52] New research and analysis indicates that including pregnant women with other groups of vulnerable persons implies that pregnant women are unable to make an informed consent,[53] and the result of this grouping is that pregnant women lose the benefits that medical research and other research provide.[54]

In a second area for debate, ethical deficiencies exist in the gap between informed consent and beneficence, and between informed consent and duty to society. In both cases (beneficence and duty to society), the literature widely agrees that informed consent may not replace the need for either of these additional elements. In other words, a research participant may not consent to participate in research that would harm himself or that would not benefit society, and it is the duty of researchers to never request such consent. This dilemma is shown in Table 2.3.

An example where informed consent and beneficence are in direct conflict can be found in the U.S. Right to Try Act of 2017, which allows terminally ill and informed patients the opportunity to knowingly request access to experimental drugs, although these drugs may cause them unknown harms. Before that law was enacted, patients could apply to the U.S. Food and Drug Administration (FDA) for access to

---

[51]   Such protected groups include, but are not exclusive to, children, prisoners (or other persons under duress), and adults without the mental capacity to make informed consent.

[52]   National Institutes of Health, "Vulnerable and Other Populations Requiring Additional Protections," web-page, updated January 7, 2019.

[53]   Indira S. E. van der Zande, Rieke van der Graaf, Martijn A. Oudijk, and Johannes J. M. van Delden, "Vulner-ability of Pregnant Women in Clinical Research," *Journal of Medical Ethics*, Vol. 43, No. 10, October 2017, pp. 657–663.

[54]   C. B. Krubiner and R. R. Faden, "Pregnant Women Should Not Be Categorised as a 'Vulnerable Population' in Biomedical Research Studies: Ending a Vicious Cycle of 'Vulnerability,'" *Journal of Medical Ethics*, Vol. 43, 2017, pp. 664–665.

unapproved drugs. The FDA reviewed existing nonpublic data and clinical trials and approved "more than 99% of requests between 2010–2015."[55]

Under the new law, patients will have access to the same drugs while circumventing the FDA's process. Opponents of the bill argued that this puts patients at unknown risks. An article in the *New England Journal of Medicine* argued that "granting very sick patients early access to unapproved products may be more likely to harm patients than to help them. Many drugs that look promising in early development are ultimately not proven safe or efficacious."[56]

Other literature considers what happens when informed consent cannot be obtained—either because the identities of affected participants are unknown at the onset of the research or because information about people is collected without their being formally enrolled as participants. Increasingly, as discussed in Chapter Five, "big data" or social media research involves the procurement and manipulation of information without explicit informed consent for each investigation. Emerging research fields that require large data sets—such as those used to train artificial intelligence algorithms—may require new ways of thinking about how to protect the interests of research subjects: people who have not consented to participate in research but who are affected nonetheless.

While informed consent may be clearly defined and documented, its implementation continues to create gray areas, such as "bystander risk," discussed in Chapter Five, where ethics can be debated by the best-informed experts.

Table 2.3
Informed Consent Versus Beneficence and Duty to Society

	. . . Beneficence	. . . Duty to Society
When informed consent conflicts with...	A research participant cannot consent to research that may cause harm to himself or herself.	A research participant cannot consent to research that may cause harm to society.
Example	A new drug is not yet approved for human trials, but patients with terminal illnesses want access to try the drug.	A couple wants to edit a gene in their embryo to protect their child from a disease, but the effect of this mutation on future descendants is unknown.

---

[55] Peter Lurie, M.D., MPH, Associate Commissioner for Public Health Strategy and Analysis, U.S. Food and Drug Administration, statement before the U.S. Senate Committee on Homeland Security and Government Affairs, September 22, 2016.

[56] S. Joffe and H. F. Lynch, "Federal Right-to-Try Legislation: Threatening the FDA's Public Health Mission," *New England Journal of Medicine*, Vol. 378, No. 8, 2018, pp. 695–697.

Integrity

Definition: *Researchers should demonstrate honesty and truthfulness. They should not fabricate data, falsify results, or omit relevant data. They should report findings fully, minimize or eliminate bias in their methods, and disclose underlying assumptions.*

The most frequently discussed aspect of integrity is the importance of honest and truthful reporting of results. These principles entail avoiding plagiarism and falsification of data and results and striving to remove bias from research methods and analysis. A great number of articles discussed risk of overrepresentation of positive results and underrepresentation of negative results in publications. The Declaration of Helsinki emphasizes that "Researchers, authors, sponsors, editors and publishers all have ethical obligations with regard to the publication and dissemination of the results of research." In publication of the results of research, both positive and negative results should be published or be publicly available. Sources of funding, institutional affiliations, and any possible conflicts of interest should be declared in the publication. Underreporting should be viewed as scientific misconduct.[57] Additionally, the rush or desire to publish leads researchers to prefer research methods that are likely to lead to surprising, or publishable, results, even if those results are not rigorous.[58]

Studies have shown that scientific studies that produced results that were statistically significant (i.e., rejected a hypothesis or otherwise did not result in the null hypothesis) or that produced positive results were more like to be published than studies that reported the null hypothesis.[59] The persistence of this publication bias, along with the pressure on researchers to publish to advance their own professional careers, can lead to dishonest actions, such as manipulating research results, on the part of the researcher.[60] This pressure and the potential for researchers to engage in dishonest actions is not unique to any one country. However, it is possible that such behaviors may be more prevalent in countries where there are fixations on numerical measures of productivity such as the quantity of publications (over quality of publication)

---

[57] World Medical Association, 2013.

[58] K. S. Button, J. P. A. Ioannidis, C. Mokrysz, B. A. Nosek, J. Flint, E. S. J. Robinson, and M. R. Munafò, "Power Failure: Why Small Sample Size Undermines the Reliability of Neuroscience," *Nature Reviews Neuroscience*, Vol. 14, No. 5, 2013, pp. 365–376.

[59] M. Malički and A. Marušić, "Is There a Solution to Publication Bias? Researchers Call for Changes in Dissemination of Clinical Research Results," *Journal of Clinical Epidemiology*, Vol. 67, No. 10, 2014, pp. 1103–1110.

[60] N. Matosin, E. Frank, M. Engel, J. S. Lum, and K. A. Newell, "Negativity Towards Negative Results: A Discussion of the Disconnect Between Scientific Worth and Scientific Culture," *Disease Models and Mechanisms*, Vol. 7, No. 2, 2014, pp. 171–173; A. Mlinarić, M. Horvat, and V. Šupak Smolčić, "Dealing with the Positive Publication Bias: Why You Should Really Publish Your Negative Results," *Biochemia Medica*, Vol. 27, No. 3, 2017.

and where researchers who are discovered engaging in fraudulent research practices encounter minimal consequences.[61]

Another important component of integrity lies in conforming to ethical rules in applying placebo and deception in research. Using a placebo as an alternative to the experimental treatment is accepted only when no current proven intervention exists or participants will not be subject to irreversible harm.[62] However, the use of deception in research is sometimes justified, such as during psychological, behavioral, and sociological studies, and must be merited such that there are no reasonable alternatives for obtaining data, and participants should not incur pain or emotional distress. In studies when deception is used, IRBs have a responsibility to consider the ramifications of its use to research participants. Participants must be accurately informed of risks and debriefed at the conclusion of the study, with the option to withdraw their data.[63]

Integrity is also important in the treatment of data. For example, data recording and data analysis should be blinded to the operator and analysts. The *British Journal of Pharmacology* states that normalization—changes to raw data—should not be undertaken unless a scientific rationale is presented.[64]

Nondiscrimination

Definition: *Researchers should minimize attempts to reduce the benefits of research on specific groups and to deny benefits from other groups.*

The principle of nondiscrimination seeks "to guarantee that human rights are exercised without discrimination of any kind based on race, colour [sic], sex, language, religion, political or other opinion, national or social origin, property, birth or other status such as disability, age, marital and family status, sexual orientation and gender identity, health status, place of residence, economic and social situation."[65] Nondis-

---

[61] A. Quin, "Fraud Scandals Sap China's Dream of Becoming a Science Superpower," *New York Times*, October 13, 2017.

[62] D. J. Harriss and G. Atkinson, "Update Ethical Standards in Sport and Exercise Science Research," *International Journal of Sports Medicine*, Vol. 32, No. 11, 2011, pp. 819–821.

[63] Harriss and Atkinson, 2011; M. A. Hall, E. Dugan, B. Zheng, and A. K. Mishra, "Trust in Physicians and Medical Institutions: What Is It, Can It Be Measured, and Does It Matter?" *Milbank Quarterly*, Vol. 79, No. 4, 2001, pp. 613–639; D. Papademas, "IVSA Code of Research Ethics and Guidelines," *Visual Studies*, Vol. 24, No. 3, 2009, pp. 250–257; APA, 2017.

[64] M. J. Curtis, R. A. Bond, D. Spina, A. Ahluwalia, S. P. A. Alexander, M. A. Giembycz, A. Gilchrist, D. Hoyer, P. A. Insel, A. A. Izzo, A. J. Lawrence, D. J. Macewan, L. D. F. Moon, S. Wonnacott, A. H. Weston, and J. C. McGrath, "Experimental Design and Analysis and Their Reporting: New Guidance for Publication in BJP," *British Journal of Pharmacology*, Vol. 172, No. 14, 2015, pp. 3461–3471.

[65] Committee on Economic, Social, and Cultural Rights, "General Comment No. 20, Non-Discrimination in Economic, Social and Cultural Rights," 2009.

crimination may apply to the population of research participants, the population affected by the research results, or the researchers themselves.

We found that when nondiscrimination is applied to research participants or the affected population, literature says that discrimination is unacceptable and that certain groups may not be excluded from research populations unless such decision is warranted by the research objectives. Even when certain groups are not excluded intentionally, there are unintentional effects of research when research participants are skewed toward one gender, race, or group. When research benefits are not equally enjoyed by all groups, discrimination may have occurred. Exceptions exist when the research itself was aimed at one particular group, such as research on tropical diseases that may disproportionately target very poor members of certain ethnic groups. That such research is sometimes described as relating to "neglected diseases" illustrates that nondiscrimination relates to what is researched as well as who participates. Nondiscrimination is violated when there are unequal distributions of the burdens or benefits of research, particularly when research is conducted on categories of patients made vulnerable by economic, social, or physical conditions and who are likely to bear only its burdens.

Dr. Rebecca Kukla, a senior research scholar at the Georgetown University Kennedy Institute of Ethics, said,

> One of the most important issues in contemporary research ethics concerns the ethical requirement that research in a specific location, involving a particular group, be culturally appropriate. On the one hand, this includes the requirement that the therapies and interventions being studied should be ones that would be welcomed by and helpful to the groups participating, and it should also be realistic that they will be available to these groups if they turn out to be effective. Availability means that they should be both affordable and realistically accessible. On the other hand, it includes the requirement that the content of the research, and the hypothesis being tested, not be damaging to or prejudiced against those participating. There is general consensus that both halves of this requirement can only be fulfilled if community members are substantively involved from the start in the formulation of research questions and methods.[66]

One approach to nondiscrimination is "fair subject selection." It is defined as the "[s]election of subjects so that stigmatized and vulnerable individuals are not targeted for risky research and the rich and socially powerful not favored for potentially beneficial research."[67] This requirement crosses our ethical principles of nondiscrimination and nonexploitation, and the latter is described further in the next section.

The APA says that "[p]sychologists recognize that fairness and justice entitle all persons to access to and benefit from the contributions of psychology and to equal

---

[66] This comment was provided to the authors by Dr. Kukla during her review of this report.

[67] Emanuel, Wendler, and Grady, 2000.

quality in the processes, procedures, and services being conducted by psychologists." Moreover, "[p]sychologists are aware of and respect cultural, individual, and role differences, including those based on age, gender, gender identity, race, ethnicity, culture, national origin, religion, sexual orientation, disability, language, and socioeconomic status, and consider these factors when working with members of such groups."[68]

New and emerging research methods try to achieve benefits without methods that lead to discrimination. One article that we encountered in our literature review examined the difference in using university students as research participants—which is a common technique across multiple disciplines when research occurs on university campuses—versus online techniques, such as crowdsourcing, to elicit research participation. The article's authors found that these newer methods lead to more-diverse groups of research participants:

> Online contract labor portals (i.e., crowdsourcing) have recently emerged as attractive alternatives to university participant pools for the purposes of collecting survey data for behavioral research. However, prior research has not provided a thorough examination of crowdsourced data for organizational psychology research. We found that, as compared with a traditional university participant pool, crowdsourcing respondents were older, were more ethnically diverse, and had more work experience. Additionally, the reliability of the data from the crowdsourcing sample was as good as or better than the corresponding university sample. Moreover, measurement invariance generally held across these groups. We conclude that the use of these labor portals is an efficient and appropriate alternative to a university participant pool, despite small differences in personality and socially desirable responding across the samples. The risks and advantages of crowdsourcing are outlined, and an overview of practical and ethical guidelines is provided.[69]

Although online pools may be more diverse than university students, they may be less diverse than the population at large.

As future research continues to be conducted online and as big data are used for disciplines such as sociology, information science, economics, and even medicine, new and emerging research methods may find ways to create more diverse *or more homogeneous* populations of research participants.[70] It will be the responsibility of researchers to examine whether their methods are unintentionally discriminatory.

When nondiscrimination is applied to the research team, the literature, including codes of conduct, encourages diversity and inclusion. The APA, for example, dictates

---

[68] APA, 2017.

[69] Behrend et al., 2011.

[70] Within the United States, social media users, on average, are younger, more highly educated, and have higher income levels than the general population. Therefore, research conducted on these platforms may exclude certain groups. See Pew Research Center, "Social Media Fact Sheet," 2018.

that "[p]sychologists exercise reasonable judgment and take precautions to ensure that their potential biases, the boundaries of their competence, and the limitations of their expertise do not lead to or condone unjust practices." In many analytic disciplines, diversity of team members has been demonstrated as a mitigation technique against biases.[71] The American Mathematical Society says that "[m]athematical ability must be respected wherever it is found, without regard to race, gender, ethnicity, age, sexual orientation, religious belief, political belief, or disability."[72] And yet, while nondiscrimination is ethical and valued, we are not aware of any examples of research being declared *unethical* because of the lack of diversity of its researchers. Therefore, diversity of the research team is, in practice, treated as aspirational and a goal to strive for, rather than an indication of ethics or lack thereof.

Nonexploitation

Definition: *Researchers should not exploit, or take unfair advantage of, research participants.*

Exploitation exists when there is unequal distribution of the burdens or benefits of research, particularly when research is conducted on categories of patients made vulnerable by impairment, institutionalization, or economic conditions and who are likely to bear only its burdens. Exploitation may occur when a population is singled out for recruitment as research participants, bears the full risks of the research, or does not enjoy the benefits of the results, and historically this includes research with racist or other prejudicial motivations. Research conducted on contraception in developing countries for the purpose of preventing people in such countries from reproducing is exploitative and violates this ethical principle. The literature suggests that nonexploitation may also be violated when a placebo is used in controlled trials when proven effective treatment is available or when intervention is likely to be used for benefit of developed countries.

The risk of exploitation increases under various circumstances, including when research is conducted outside the proximity of IRBs or strict legal frameworks, such as in developing countries. In "Ethical and Scientific Implications of the Globalization of Clinical Research" in the *New England Journal of Medicine*, doctors and researchers write about the benefits and risks of conducting medical research in developing countries:

> There are clear benefits to conducting trials in developing countries. These include fostering positive relationships among clinician investigators globally and answer-

---

[71] H. Valantine, "The Science of Diversity and the Impact of Implicit Bias," National Institutes of Health, 2017.

[72] American Mathematical Society, "Policy Statement on Ethical Guidelines," 2005.

ing questions about the safety and efficacy of drugs and devices that are of interest throughout the world. At the same time, the globalization of clinical trials raises ethical and scientific concerns. . . . Wide disparities in education, economic and social standing, and health care systems may jeopardize the rights of research participants. There may be a relative lack of understanding of both the investigational nature of therapeutic products and the use of placebo groups. In some places, financial compensation for research participation may exceed participants' annual wages, and participation in a clinical trial may provide the only access to care for persons with the condition under study. Standards of health care in developing countries may also allow ethically problematic study designs or trials that would not be allowed in wealthier countries. In one study, only 56% of the 670 researchers surveyed in developing countries reported that their research had been reviewed by a local institutional review board or health ministry. In another study, 90% of published clinical trials conducted in China in 2004 did not report ethical review of the protocol and only 18% adequately discussed informed consent.[73]

This literature implies that perhaps researchers in developing countries do not always follow the same ethics protocols—such as review by an IRB or use of informed consent—as they would in the United States. Doing so violates the principle of nonexploitation. The authors indicate that as of 2009, research conducted in developing countries may exploit the local population, who cannot reasonably be expected to benefit from the research when the resulting medical advances are too expensive. Thus, adherence to nonexploitation requires that research not exploit one population to benefit another, and examples of ethical violations and adherence can be found from the disciplines of information science to medicine.[74]

One of our interview participants emphasized this point at length, saying, "there are reports of researchers acquiring genetic data without the consent of the individuals because the consent is not required in some countries."[75] We asked whether researchers were ethically justified in conducting research outside of their home country to reduce research costs or produce results that may help humanity sooner by cutting through bureaucratic red tape. Our interviewee was unwavering: Such regulations exist to protect research participants, and circumventing them by conducting research elsewhere exploits the local population where the research is conducted. In this expert's experience, pharmaceutical trials are conducted in countries with fewer regulations as "common practice," and this is ethically concerning.[76] This same interviewee chal-

---

[73] S. W. Glickman, J. G. McHutchinson, E. D. Peterson, C. B. Cairns, R. A. Harrington, R. M. Califf, and K. A. Schulman, "Ethical and Scientific Implications of the Globalization of Clinical Research," *New England Journal of Medicine*, Vol. 360, No. 8, 2009, pp. 816–823.

[74] Glickman et al., 2009.

[75] Interview 5.

[76] Interview 5.

lenged the view that such decisions are made based on cost savings, at least in situations where the entire study team and laboratory equipment need to be relocated.

One case study involving the experience of the San people, an indigenous people of Africa, illustrates the potential to empower potentially vulnerable populations of interest to researchers.[77]

> Four San individuals, the eldest in their respective communities, were chosen for genome sequencing, and the published article analysed many aspects of the correlations, differences and relationships found in the single-nucleotide polymorphisms (SNPs). . . . A supplementary document published with the paper contained numerous conclusions and details that the San regarded as private, pejorative, discriminatory and inappropriate. The San leadership met with the authors in Namibia soon after publication, asking why they as leaders had not been approached for permission in advance, and enquiring about the informed consent process. The authors refused to provide details about the informed consent process, apart from stating that they had received video-recorded consents in each case. They defended their denial of the right of the San leadership to further information on the grounds that the research project had been fully approved by ethics committees/institutional review boards in three countries, . . . and that they had complied with all the relevant requirements.[78]

This research on the San people began in 2009 and was published in 2010. In 2017, the San people became the first indigenous group known to create their own code of research ethics. It includes statements such as the following:

> We require respect, not only for individuals but also for the community. . . . We require respect for our culture, which also includes our history. We have certain sensitivities that are not known by others. Respect is shown when we can input into all research endeavours at all stages so that we can explain these sensitivities. . . . We require honesty from all those who come to us with research proposals.[79]

---

[77] Doris Schroeder, J. C. Cook, Francois Hirsch, Solveig Fenet, and Vasantha Muthuswamy, *Ethics Dumping, Case Studies from North-South Research Collaborations*, New York: Springer International, 2017; Linda Nordling, "San People of Africa Draft Code of Ethics for Researchers," *Science*, March 17, 2017.

[78] R. Chennells and Andries Steenkamp, "International Genomics Research Involving the San People," in Doris Schroeder, J. C. Cook, Francois Hirsch, Solveig Fenet, and Vasantha Muthuswamy, *Ethics Dumping, Case Studies from North-South Research Collaborations*, New York: Springer International, 2017.

[79] South African San Institute and the TRUST Project, "San Code of Research Ethics," Kimberley, South Africa, 2017.

Privacy and Confidentiality

Privacy Definition: *Research participants have the right to control access to their personal information and to their bodies in the collection of biological specimens. Participants may control how others see, touch, or obtain their information.*

Confidentiality Definition: *Researchers will protect the private information provided by participants from release. Confidentiality is an extension of the concept of privacy; it refers to the participant's understanding of, and agreement to, the ways identifiable information will be stored and shared.*

According to the Declaration of Helsinki, "[e]very precaution must be taken to protect the privacy of research subjects and the confidentiality of their personal information."[80] Privacy and confidentiality apply to research that uses human participants or data about humans.

The privacy issues raised in the literature center on the management of research participants' information. It begins with the protocols that the scientific community should follow to ensure against the disclosure of personal or confidential information. These include de-identifying personal data, encrypting it (along with the codes used to link identities), limiting access to a minimum number of people, and planning for how confidentiality will be maintained when information is shared among sponsors, collaborators, or coinvestigators.[81]

In regard to the management of personal information, absolute secrecy is not expected; instead, limited disclosure is the preferred practice. In other words, personal data can be shared among researchers who need access to the information, but not outside of a trusted group.[82] The literature reminds researchers and practitioners that they must be cognizant of the potential personal or economic repercussions that may emanate from the disclosure of personal data. The unconsented disclosure of information can take place if all of the following conditions are simultaneously satisfied:

- The information that has been collected is important.[83]
- Consent is difficult or impossible to obtain.
- Objection by a reasonable individual to publication seems unlikely.
- The identity of the source of information or data is protected.[84]

---

[80] WMA, 2013.

[81] Chan et al., 2013.

[82] Hall et al., 2001.

[83] The literature did not define how "important" is measured, leaving this to the researcher's judgment and/or the review board.

[84] F. Portaluppi, M. H. Smolensky, and Y. Touitou, "Ethics and Methods for Biological Rhythm Research on Animals and Human Beings," *Chronobiology International*, Vol. 27, 9–10, 2010, pp. 1911–1929.

A growing issue in research is the use of publicly available data, or quasi publicly available data (such as social media data available only to one's "friends"), for research. The *privacy paradox* describes the situation whereby consumers express high levels of interest in protecting their personal information, yet freely give it away in certain circumstances. Such behavior has been explained by the transformation of the privacy construct from a civil right to a commodity used by consumers as a means of exchange for utility or economic gain. A second issue is an examination of the degrees to which a person's identity can be concealed.[85] The implications of big data for research, including social media data, are discussed in Chapter Five.

Several articles and codes of conduct describe protocols for securing data—such as when to use encryption or secured storage techniques and other methods—but these discussions require that researchers have already decided which data and analysis to protect.[86] In the United States, this decision may be clearest when dealing with types of information covered by law or regulation, such as the health-related information protected by the Health Information Portability and Accountability Act, the genetic information protected by the Genetic Information Nondiscrimination Act (GINA), the financial information protected by the Gramm-Leach Bliley Act, and so on.

The ACM highlights the special responsibilities of computer scientists and related professionals because of their understanding of, and role in designing, technologies for collecting, storing, analyzing, and communicating:

> The responsibility of respecting privacy applies to computing professionals in a particularly profound way. Technology enables the collection, monitoring, and exchange of personal information quickly, inexpensively, and often without the knowledge of the people affected. Therefore, a computing professional should become conversant in the various definitions and forms of privacy and should understand the rights and responsibilities associated with the collection and use of personal information. Computing professionals should only use personal information for legitimate ends and without violating the rights of individuals and groups. This requires taking precautions to prevent re-identification of anonymized data or unauthorized data collection, ensuring the accuracy of data, understanding the provenance of the data, and protecting it from unauthorized access and accidental disclosure. Computing professionals should establish transparent policies and procedures that allow individuals to understand what data is being collected and how it is being used, to give informed consent for automatic data collection, and to review, obtain, correct inaccuracies in, and delete their personal data. Only the minimum amount of personal information necessary should be collected in a system. The retention and disposal periods for that information should be clearly

---

85 H. J. Smith, T. Dinev, and H. Xu, "Information Privacy Research: An Interdisciplinary Review," *MIS Quarterly: Management Information Systems*, Vol. 35, No. 4, 2011, pp. 989–1015.

86 Researchers also need to keep track of changes in the effectiveness of any particular mechanism—what works today may not work tomorrow.

defined, enforced, and communicated to data subjects. Personal information gathered for a specific purpose should not be used for other purposes without the person's consent. Merged data collections can compromise privacy features present in the original collections. Therefore, computing professionals should take special care for privacy when merging data collections.[87]

Aligned with principles set forth in the Declaration of Helsinki, codes of conduct instruct researchers to safeguard any information given to them in confidence. Across codes of conduct, professional societies and associations detail how privacy and confidentiality should be maintained within their discipline. The American Society of Human Genetics, for example, asks that its members seek the consent of a patient when disclosing the patient's data or consider "legal, ethical, and professional" obligations when faced with a scenario where such considerations are necessary. The American Medical Association (AMA) asks that its members make a point of clarifying their consent processes so that patients understand the privacy standards to which their data will be subject, including whether it will be encrypted or "completely de-identified." The APA instructs its members to take reasonable steps to protect any information obtained from patients based on professional standards or relevant regulations. APA has also established guidelines for minimizing intrusions to patient privacy and how to deal with disclosures.

The Academy of Management stresses that all confidential information must be protected, even when there is no legal basis for it. This entails eliminating identifier information if it is made available to the public and accounting for long-term use if it were ever examined by other stakeholders or published in the public record. The bodies within this discipline emphasize the relationship of researchers and practitioners. In many cases (i.e., American Anthropological Association, International Sociological Association, Society for Applied Anthropology, and IVSA), there is a duty to not disclose information, de-identify any data, and clearly explain to participants the realistic limits of confidentiality. See Box 2.2 for a discussion of the evolution of privacy guidance both in the United States and internationally.

Privacy and confidentiality can extend beyond data on persons to include intellectual property and other proprietary information. According to the Engineering Codes of Ethics, "[e]ngineers in all areas of professional practice frequently become privy to information that is intended by the employer or client to remain confidential. It may be sensitive employer or client information, trade secrets, technical processes, or business information that, if disclosed or used improperly, could damage the business or other interests of the employer or client."[88]

---

[87] ACM, 2018.

[88] A. E. Schwartz, "Engineering Society Codes of Ethics: A Bird's-Eye View," *The Bridge*, Vol. 47, No. 1, 2017, pp. 21–26.

Box 2.2
Broad Privacy Policy Affecting Research and Beyond

Privacy policy got its start in the United States, and its more-recent evolution has occurred internationally, especially in Europe. The circumstances surrounding the Belmont Report were complemented by those resulting in the Privacy Act of 1974, which grew out of the 1973 *Records, Computers, and the Rights of Citizens* report of the former U.S. Department of Health, Education, and Welfare. That report articulated fair information practice principles that have had global influence on public policy, commercial practice, and research. These principles evolved through the multilateral Organisation for Economic Co-operation and Development (OECD, members of which are industrialized countries, including the United States and countries around the world) beginning in 1980 and most recently updated in 2013. The OECD's codification of privacy principles has, in turn, influenced individual country and European Union (EU) policy on privacy, beginning with the 1995 Data Protection Directive and extending to the 2018 General Data Protection Regulation.

Privacy principles associated with this history connect to principles discussed here under Privacy and Confidentiality, Informed Consent, and Integrity, in particular.

SOURCE: Electronic Privacy Information Center, "The Privacy Act of 1974," 2018.

Protection of privacy can be in tension with other ethical concerns. One is the value of sharing data for research transparency and reproducibility (see discussion of Open Science in Chapter Five). Another is the new concern that broader and more-thorough collection of data about groups and populations could compromise the interests of those groups or populations—a variation on the concern for the individual traditionally associated with privacy but relevant to considerations of confidentiality. As emerging fields collect more society-wide data, will the private information of an entire society be revealed if data are compromised?

Professional Competence

Definition: *Researchers should engage only in work that they are qualified to perform, while also participating in training and betterment programs with the intent of improving their skill sets. This principle includes choosing appropriate research methods, statistical methods, and sample sizes to avoid misleading results.*

Across disciplines, professional competence is described as an ethical principle that presupposes that researchers are trained in and using appropriate research methods. Additionally, this ethical principle suggests that researchers adhere to appropriate safety standards when conducting their research. Professional competence places the responsibility on the researchers to have such knowledge, training, and awareness. It does not allow ignorance of certain research methods or research practices due to lack of training or awareness to act as a justification for noncompliance. Recent growth in noncredentialled individuals engaging in research is discussed in Chapter Five.

With respect to choosing the appropriate research method, Douglas G. Altman, who wrote hundreds of articles on statistical analysis, explained professional competence in this way:

What should we think about a doctor who uses the wrong treatment, either wilfully [sic] or through ignorance, or who uses the right treatment wrongly (such as by giving the wrong dose of a drug)? Most people would agree that such behavior was unprofessional, arguably unethical, and certainly unacceptable. What, then, should we think about researchers who use the wrong techniques (either wilfully [sic] or in ignorance), use the right techniques wrongly, misinterpret their results, report their results selectively, cite the literature selectively, and draw unjustified conclusions? We should be appalled. Yet numerous studies of the medical literature, in both general and specialist journals, have shown that all of the above phenomena are common.[89]

To Altman and many of the professional disciplines we examined, the wrong selection of research methods, the wrong use of a control group, and the wrong selection of sampling are as unethical as other behaviors that lead to deceitful research results or put research participants at unnecessary risk. He told researchers to consider their sample size,[90] while additional authors have advised researchers to understand and use specific methods, like Delphi,[91] randomization,[92] and types of control groups.[93]

The literature we examined focused less on the willful neglect of research participants and more on how ill-advised or uninformed decisions by researchers could lead to the same unethical result. In neuroscience, for example, an article from 2013 argues that the low sample size of many research projects in the field creates misleading results, or simply, "a reduced chance of detecting a true effect." In this and like research fields, the authors argue, this challenge is significant because many studies have trouble finding large numbers of research participants who meet the study's selection criteria.

The challenges of professional competence can be complicated where defining the profession is itself a challenge. As new cross-disciplinary fields emerge, researchers might be challenged to know whether they have brought the right expertise and competence onto their team. For example, in many modern and emerging scientific disciplines—such as the proliferating "omics" fields (genomics, proteomics, metabolomics, and so on), where researchers discuss the need for knowledge that crosses disciplines

---

[89] D. G. Altman, "The Scandal of Poor Medical Research: We Need Less Research, Better Research, and Research Done for the Right Reasons," *British Medical Journal*, Vol. 308, No. 6924, 1994, pp. 283–284.

[90] D. G. Altman, "Statistics and Ethics in Medical Research, III: How Large a Sample? *British Medical Journal*, Vol. 281, No. 6251, 1980, pp. 1336–1338.

[91] F. Hasson, S. Keeney, and H. McKenna, "Research Guidelines for the Delphi Survey Technique," *Journal of Advanced Nursing*, Vol. 32, No. 4, 2000, pp. 1008–1015.

[92] D. G. Altman, K. F. Schulz, D. Moher, M. Egger, F. Davidoff, D. Elbourne, P.C. Gøtzsche, and T. Lang, "The Revised CONSORT Statement for Reporting Randomized Trials: Explanation and Elaboration," *Annals of Internal Medicine*, Vol. 134, No. 8, 2001, pp. 663–694.

[93] R. B. D'Agostino, Sr., J. M. Massaro, and L. M. Sullivan, "Non-Inferiority Trials: Design Concepts and Issues—The Encounters of Academic Consultants in Statistics," *Statistics in Medicine*, Vol. 22, No. 2, 2003, pp. 169–186.

and for collaborations to advance these fields—expectations are changing both for the breadth of an individual researcher's knowledge and for the breadth of expertise in a team, including

> the skills to understand biological systems and to use that information effectively for the benefit of humankind. . . . As biomedical research is becoming increasingly data intensive, computational capability is increasingly becoming a critical skill. Although a good start has been made, expanded interactions will be required between the sciences (biology, computer science, physics, mathematics, statistics, chemistry and engineering), between the basic and the clinical sciences, and between the life sciences, the social sciences and the humanities. Such interactions will be needed at the individual level (scientists, clinicians and scholars will need to be able to bring relevant issues, concerns and capabilities from different disciplines to bear on their specific research efforts), at a collaborative level (researchers will need to be able to participate effectively in interdisciplinary research collaborations that bring biology together with many other disciplines) and at the disciplinary level (new disciplines will need to emerge at the interfaces between the traditional disciplines).[94]

To accomplish this goal, the authors describe the value of intellectual diversity or different perspectives:

> Individuals from minority or disadvantaged populations are significantly underrepresented as both researchers and participants in genomics research. This regrettable circumstance deprives the field of the best and brightest from all backgrounds, narrows the field of questions asked, can lessen sensitivity to cultural concerns in implementing research protocols, and compromises the overall effectiveness of the research.[95]

An article in the *Australian and New Zealand Journal of Psychiatry* provides a table of key issues to consider when evaluating the quality of qualitative research design. Table 2.4 displays a modified version of that table, which provides researchers with criteria and considerations for choosing the appropriate research methods.

Within actual codes of conduct, professional competence is documented and described in similar ways. The AMA says researchers should "participate only in those studies for which they have relevant expertise" and "assure themselves that the research protocol is scientifically sound and meets ethical guidelines for research with human participants."[96] The National Academies say researchers are responsible for remain-

---

[94] Collins et al., 2003.

[95] Collins et al., 2003.

[96] AMA, "Code of Medical Ethics of the American Medical Association: Opinions on Research and Innovation," 2016.

Table 2.4
Considerations for Research Methods

Criteria	Considerations
Congruence	• Do the methods used "fit" with the chosen methodology? • Is the study conducted in a way that is congruent with the stated methodology (i.e., philosophical/theoretical approach)?
Responsiveness to social context	• Was the research design developed and adapted to respond to real-life situations within the social settings in which it was conducted? • Did the researchers engage with participants and become familiar with the study context?
Appropriateness	• Were the sampling strategies suitable to identify participants and sources to inform the research question being addressed? • Were suitable data gathering methods used to inform the research question being addressed?
Adequacy	• Have sufficient sources of information been sampled to develop a full description of the issue being studied? • Is there a detailed description of the data gathering and analytical processes followed? • Were multiple methods and/or sources of information weighed in the analysis? • Were methods of gathering and recording/documenting data sensitive to participants' language and views? • Were corroborating, illuminating, and rival accounts gathered and analyzed to explore multiple aspects of the research issue?
Transparency	• To what extent have the processes of data gathering and analysis been rendered transparent? • How were rival/competing accounts dealt with in the analysis?
Authenticity	• Are participants' views presented in their own voices—that is, are verbatim quotes presented? • Are a range of voices and views (including dissenting views) represented? • Would the descriptions and interpretations of data be recognizable to those having the experiences/in the situations described?
Coherence	• Do the findings "fit" the data from which they are derived—that is, are the linkages between data and findings plausible? • Have the perspectives of multiple researchers (e.g., research team) been taken into account—e.g., are corroborating and competing elements considered?
Reciprocity	• To what extent were processes of conducting/reviewing the analysis/negotiating the interpretations shared with participants?
Typicality	• What claims are made for generalizability of the findings to other bodies of knowledge, populations, or contexts/settings?
Permeability of the researcher's intentions, engagement, interpretations	• Did the study develop/change the researcher's initial understanding of the social worlds/phenomena studied? • Are the researcher's intentions, preconceptions, values, or preferred theories revealed in the report?

SOURCE: Adapted from E. Fossey, C. Harvey, F. McDermott, and L. Davidson, "Understanding and Evaluating Qualitative Research," *Australian and New Zealand Journal of Psychiatry*, Vol. 36, No. 6, 2002, pp. 717–732.

ing knowledgeable on current techniques and methods and for reporting results "as objectively and as accurately as possible."[97] And the ASM says, "ASM members strive to increase the competence and prestige of the profession and practice of microbiology by responsible action."[98]

The American Psychological Association discusses how to handle professional competence in a new emerging area: "In those emerging areas in which generally recognized standards for preparatory training do not yet exist, psychologists nevertheless take reasonable steps to ensure the competence of their work and to protect clients/patients, students, supervisees, research participants, organizational clients, and others from harm."[99]

The British Computing Society says researchers shall "accept professional responsibility for your work and for the work of colleagues who are defined in a given context as working under your supervision" and shall

> a. only undertake to do work or provide a service that is within your professional competence.
>
> b. NOT claim any level of competence that you do not possess.
>
> c. develop your professional knowledge, skills and competence on a continuing basis, maintaining awareness of technological developments, procedures, and standards that are relevant to your field.
>
> d. ensure that you have the knowledge and understanding of Legislation and that you comply with such Legislation, in carrying out your professional responsibilities.
>
> e. respect and value alternative viewpoints and seek, accept and offer honest criticisms of work.
>
> f. avoid injuring others, their property, reputation, or employment by false or malicious or negligent action or inaction.
>
> g. reject and will not make any offer of bribery or unethical inducement.[100]

Codes of conduct state that "research plans and protocols should . . . demonstrate that the study design has the critical elements . . .";[101] "under no circumstance should environmental epidemiologists engage in selecting methods or practices that are

---

[97] National Academy of Sciences, National Academy of Engineering, and Institute of Medicine of the National Academies, 2009.

[98] ASM, 2005.

[99] APA, 2017.

[100] British Computing Society, "Code of Conduct for BCS Members," 2015.

[101] International Society for Environmental Epidemiology, 2012.

designed to produce misleading results, nor should they misrepresent findings,"[102] and "Ecologists will offer professional advice and guidance only on those subjects in which they are informed and qualified through professional training or experience. They will strive to accurately represent ecological understanding and knowledge and to avoid and discourage dissemination of erroneous, biased, or exaggerated statements about ecology."[103] In the American Geophysical Union, "Members will employ research methods to the best of their understanding and ability, base conclusions on critical analysis of the evidence, and report findings and interpretations fully, accurately, and objectively, including characterization of uncertainties."[104]

In the American Statistical Association,

> The ethical statistician uses methodology and data that are relevant and appropriate; without favoritism or prejudice; and in a manner intended to produce valid, interpretable, and reproducible results. The ethical statistician does not knowingly accept work for which he/she is not sufficiently qualified, is honest with the client about any limitation of expertise, and consults other statisticians when necessary or in doubt.[105]

We found no evidence that various disciplines disagree about professional competence. Rather, when differences emerge, they tend to be along either (a) issues that are specific to that discipline and do not normally arise in other disciplines, or (b) issues that other disciplines would probably agree with but possibly haven't addressed yet.[106]

Professional Discipline

Definition: *Researchers should engage in ethical research and help other researchers engage in ethical research by promoting ethical behaviors through practice, publishing and communicating, mentoring and teaching, and other activities.*

Professional discipline relates to how a researcher adheres to ethics; how a researcher promotes ethics, including through mentoring and training other research-

---

[102] International Society for Environmental Epidemiology, 2012.

[103] Ecological Society of America, "ESA Code of Ethics," 2013.

[104] American Geophysical Union, "AGU Scientific Integrity and Professional Ethics," 2017.

[105] American Statistical Association, "Ethical Guidelines for Statistical Practice," 2018.

[106] An example of the first point is from the International Society of Ethnobiology, which requires researchers to understand the local context prior to entering into research relationships with a community and to conduct research in the local language, which was not stated in any other literature from our search. An example of the second point is from the American Statistical Association, which requires researchers to use "methodology and data that are relevant and appropriate; without favoritism or prejudice; and in a manner intended to produce valid, interpretable, and reproducible results."

ers and acting as a reviewer for other researchers' studies; and how a researcher enforces ethics, including by conducting peer review of research submitted through publication and other activities. Professional discipline implies the internalization of ethical principles and their external expression in behavior across the board. It requires researchers to do their research and related activities ethically, and it encourages sponsoring agencies and professional outlets, such as societies or journals, to enforce ethical practice. In some cases, codes of conduct differentiate between professional discipline (researchers should promote ethical practice within their discipline) and adherence to code (researchers should themselves practice ethically). In this report, we combine principles.

Examples of professional discipline appear in connection with researchers reviewing each other's results. The National Academies say, "reviewers and readers of scientific papers" have a responsibility "to evaluate not only the validity of the data but also the reliability of the methods used to derive those data." They observe that while honest errors will occur, errors caused by negligence—including haste, carelessness, and inattention—can cause "serious damage both within science and in the broader society that relies on scientific results." Achieving the recommended evaluation implies access to data and to methods, which may imply access to computer code (see the discussion of Open Science in Chapter Five).

The AMA requires all physicians to "uphold the standards of professionalism, be honest in all professional interactions, and strive to report physicians deficient in character or competence, or engaging in fraud or deception, to appropriate entities."[107] The American Society of Human Genetics says that members should "[p]romote the health of the public, through the advancement of human genetic research and the provision of high quality genetic services conducted to the highest ethical and professional standards." And, lastly, a noteworthy description of professional discipline comes from the IVSA:

> Visual researchers adhere to the highest professional standards and accept responsibility for their work. Members of IVSA understand that they form a community and show respect for others even when they disagree on theoretical, methodological, or personal approaches to visual research, which also places value on the public trust in research activity, demarcating it from other potentially disreputable visual practices. The professional and public trust rests on the ethical behavior of people doing ethical visual research. IVSA is vigilant to separate ethical visual practices from those that intentionally violate that trust. For this IVSA represents a shared responsibility for ethical research.[108]

---

[107] AMA, 2018.

[108] Papademas, 2009.

Professional discipline may be the context in which harassment, a behavioral concern that has always existed, is generating renewed attention and more-vigorous action. People we interviewed involved with journal publication and with ethics in research broadly noted that the research community is paying more attention to how researchers treat each other. Harassment can discourage whistle-blowing and practice of ethical conduct, and it can also discourage continued engagement in research by its victims.

CHAPTER THREE

International Landscape of Ethics

How ethics are codified differs from country to country just as from discipline to discipline. This chapter examines how these international differences manifest in both literature and practice. We examined literature on research ethics from Europe, China, Russia, Africa, and global organizations. We also interviewed experts who conduct research across international borders or who work with researchers from various countries.

Our first finding is that a distinction exists between research conducted in any particular country with researchers from that country versus research conducted with researchers from *other* countries. This distinction is important in understanding whether ethical differences are a result of local customs, culture, laws, and practices or result from one culture being subjected to the ethics of foreigners.

In the first case, local culture and norms shape laws, regulations, and codes of ethics, and just as culture and norms differ from country to country, so do the laws, regulations, and codes of ethics that they inspire. The Declaration of Helsinki advocates a universal view of ethics, which we discuss in detail below. Then that universal view meets the realities of local attitudes and practices, where cultures differ on topics such as whose consent is important (e.g., the individual versus the family, tribe, and community) and other differences in cultural values. We found that there may also be differences in knowledge, education, and training from region to region, which may lead to differences in understanding of research safety and ethical practices.

In the second case, the increasing ease of travel and communication results in research sometimes being conducted in locations other than the researchers' home countries or in the country where the research is funded, analyzed, and published. Remote research predates Charles Darwin, yet new technologies and other factors make it easier to do now, just as collaboration with researchers in other countries is easier and increasingly common. Researchers can now use global "big data" data sets involving research participants from 100 countries or more in a single study. Pharmaceutical and health care companies have long conducted research in foreign countries, while mapping the human genome involved researchers around the world using information technology.

U.S. government funders of research, including the Department of Health and Human Services (HHS) and the National Science Foundation, have long supported research undertaken outside of the United States, with the expectation that researchers engage in the same ethics as they would follow in the United States. U.S. human subjects' protection committees or IRBs are expected to know about research contexts (including law and culture) in countries where the research will take place, including what specimens and data can be exported.

This chapter discusses how international ethics have been created and instantiated, where some differences occur, and lessons that have been learned when cultures conflict. It focuses on the implications for ethics in research being conducted in contrasting circumstances around the world, not through global collaborations, per se. The discussion focuses further on ethics associated with how research is conducted. When researchers undertake their work in a different community (whether in their own country or another), the question of what kind of research they do also matters— ethical research is culturally appropriate. For that to be the case implies input from the community of affected research participants into research planning, which anecdotally remains more of an aspiration than a common condition.

The Seminal Influence of the Declaration of Helsinki

The Declaration of Helsinki, named for the location where it was first adopted in 1964 by the WMA, became "the first international set of ethical principles for research involving human subjects."[1] Developed from the Nuremberg Code, it was approved unanimously by national medical associations composing the WMA[2] and is the starting point for many discussions of and documents about international ethics.[3]

The Declaration "is intended to be read as a whole, and each of its constituent paragraphs should be applied with consideration of all other relevant paragraphs."[4] Its sections discuss informed consent, privacy and confidentiality, attention to ethics in research protocols and research oversight by committees concerned with research ethics, publication of research results, minimization of risks to participants (and the expectation that potential benefits outweigh the risks), specific consideration of implications for vulnerable groups, avoidance of harm to the environment, and other topics.

The Declaration requires physicians to "protect the life, health, dignity, integrity, right to self-determination, privacy, and confidentiality of personal information

---

[1]   U. Wiesing, "The Declaration of Helsinki—Its History and Its Future," World Medical Association, November 11, 2014.

[2]   Wiesing, 2014.

[3]   The Declaration was amended several times between 1975 and 2013 (WMA, 2013).

[4]   WMA, 2013.

of research subjects" because "[i]t is the duty of the physician to promote and safe-guard the health, well-being and rights of patients, including those who are involved in medical research."[5] It requires physicians to consider both their own national and international ethical, legal, and regulatory norms, none of which should abridge the protection in the Declaration. These same ethical principles and concepts for professional conduct can be found today in many of the codes of conduct we reviewed, even in scientific disciplines not related to medicine.

The Declaration arose in response to unethical experiments conducted by the Nazis and to the post–World War II formation of the WMA.[6] After first developing an update to the Hippocratic Oath as the Declaration of Geneva, the WMA's consideration of a report on medicine and war crimes led to an international code of medical ethics in 1949 and the establishment of a standing Committee on Medical Ethics in 1952.[7] The adoption by the WMA of the Declaration of Helsinki in 1964 set in motion a process of evolving ethical principles premised on the value of human-subjects research in medicine at periodic WMA general assemblies held in different countries. The WMA today includes 113 constituent members (i.e., national associations of physicians).[8] Its annual report documents concern about and interventions to alleviate practices in different countries.[9] The current version of the Declaration was adopted in 2013.

The Convention on Biological Diversity and Its Derivatives

Whereas the Helsinki Declaration focuses on biomedical research, other international agreements have focused on the environment and ecology. Perhaps most prominent is the Convention on Biological Diversity (CBD),[10] which was developed between the late 1980s and the early 1990s to balance interests in sustainable development, conservation and avoidance of species extinction, and fair and equitable sharing of benefits from genetic materials (i.e., plants, animals, and microorganisms), especially materials collected in economically challenged nations for uses based elsewhere.[11] The CBD and

---

[5] WMA, 2013.

[6] WMA, "History—The Story of the WMA," 2018.

[7] WMA, 2018. The WMA also publishes the WMA Medical Ethics Manual, now available in 23 languages (WMA, Medical Ethics Manual, 2005).

[8] WMA, 2018. Members include national associations of physicians (constituent members) or individual physicians (associate members).

[9] WMA, Annual Report, 2017.

[10] CBD, "The Cartagena Protocol on Biosafety," 2003.

[11] CBD, "History of the Convention," webpage, 2018a.

Box 3.1
Convention on Biological Diversity and Its Supplements
- Convention on Biological Diversity (1993) says benefits to humans should not be at the expense of biodiversity and sustainability of the ecosystem.
- Cartagena Protocol on Biosafety (2003) addresses risks to people and the ecosystem from the handling of live organisms associated with biotechnology.
- Nagoya Protocol on Access and Benefit Sharing (2014) provides a legal framework for transnational use of genetic material and is associated with an information clearinghouse.

its associated agreements are listed in Box 3.1. They share an integrated website linking them to "safeguarding life on earth."

This set of agreements could be seen as expressions of "duty to society" associated with environmental stewardship in its fullest sense—connecting human health and well-being to the whole ecosystem in which people live:

> [T]he Convention recognizes that biological diversity is about more than plants, animals and micro organisms and their ecosystems—it is about people and our need for food security, medicines, fresh air and water, shelter, and a clean and healthy environment in which to live.[12]

It has particular bearing on research that involves collecting samples or genetically modifying organisms in one area for analysis and/or use elsewhere.

The CBD's interest-balancing implies that benefiting humans should not be at the expense of biodiversity, and the sustainable development emphasis reinforces an ecosystem orientation. The United Nations (UN), which oversees the CBD Secretariat, takes an expansive view of the impact of the CBD: "In fact, it covers all possible domains that are directly or indirectly related to biodiversity and its role in development, ranging from science, politics and education to agriculture, business, culture and much more."[13]

The Precautionary Principle

The CBD complements the 1998 articulation of the Precautionary Principle at the Wingspread Conference of U.S., European, and Canadian scientists, philosophers, lawyers, and environmental activists.[14] The Precautionary Principle calls for a risk-averse approach to decisionmaking on public health and environmental concerns, specifically calling for action if harm is anticipated but scientific uncertainty is significant.

---

[12] CBD, "The Convention on Biological Diversity," webpage, 2018b.

[13] UN, "Convention on Biodiversity," webpage, 2018.

[14] Science and Environmental Health Network, "Wingspread Conference on the Precautionary Principle," webpage, 1998.

In that spirit, the UN CBD website asserts that risk to biodiversity should override concerns about scientific uncertainty: "The precautionary principle states that where there is a threat of significant reduction or loss of biological diversity, lack of full scientific certainty should not be used as a reason for postponing measures to avoid or minimize such a threat."[15] Its intrinsic risk-aversion, which has implications for public policy (e.g., it has been embraced by EU policy) as well as for research, is at the heart of debates over the suitability of the Precautionary Principle in different contexts.[16] The Precautionary Principle motivates forbearance, a better-safe-than-sorry approach to decisionmaking about research and practice that can chill exploration and innovation. How to adapt the Precautionary Principle (or whether that is even feasible) as new fields emerge and research horizons change is an enduring challenge.

Regional Differences

Different views of research ethics around the world abound, reflecting differences in culture.[17] One area where this is evident in the literature is to what extent or degree a culture values the individual over society or the family or community over the individual. Even where there are high-level conceptual commonalities, specific differences are reflected in how countries define research, human subjects, and privacy-sensitive information.[18] HHS houses the Office for Human Research Protections (OHRP), which both provides guidance and oversight for U.S. human subjects protection activities and monitors similar activities around the world.[19] Its clustering of research guidance of different kinds into various categories (general; drugs, biologics, and devices; clinical trials registries; research injury; social-behavioral research; privacy and data protection; human biological materials; genetics research; and embryos, stem cells, and cloning) provides a way of demonstrating significant commonality as well as indicating which countries have a lot of guidance and which have less in each year the guidance is compiled.[20]

---

[15] UN, 2018.

[16] D. Kriebel, J. Tickner, P. Epstein, J. Lemons, R. Levins, E. L. Loechler, and M. Stoto, "The Precautionary Principle in Environmental Science," *Environmental Health Perspectives*, Vol. 109, No. 9, 2001, pp. 871–876; EU, The Precautionary Principle, Communication(2000) 1Final, February 2, 2000; K. Garnett and D. J. Parsons, "Multi-Case Review of the Application of the Precautionary Principle in European Union Law and Case Law," *Risk Analysis*, Vol. 37, No. 3, 2017, pp. 502–516.

[17] Interview 11.

[18] Interview 3.

[19] For a comprehensive listing of laws, regulations, and guidelines for research involving human subjects across countries, see HHS, "International Compilation of Human Research Standards," 2018.

[20] HHS, "International Compilation of Human Research Standards," 2018.

Many European countries, for example, do not define research in their codes of ethics (unlike HHS), and they may differ in how they define human subjects (for example, someone who is deceased is not considered a human subject in the United States, but in some European countries a deceased person would be a human subject for one year post mortem).[21] The new European General Data Protection Regulation has a very expansive definition of personal information that may warrant protection, whereas in the United States, there is a narrower (and often domain-specific) characterization of privacy-sensitive information. In some countries, a participant is someone who gives consent by enrolling in research (e.g., India); in others, it may be someone whose data or responses are relevant to answering research questions (e.g., this was true until 2014 for Canada).[22] See Box 3.2 for comparisons on informed consent.

Europe presents an interesting situation because multiple countries with different histories, cultures, and values have agreed, at least in principle, to share certain policies and programs through their membership in the EU. When it comes to research, the EU has scale that individual member countries lack, and it accompanies broad research initiatives and funding with research policy promulgated by the EC.[23] The expected benefits of scale combine with an express linkage to ethics in the framing of a new AI4EU initiative,[24] which calls for an associated "ethical observatory." Broadly speaking, the EC has a highly regulatory approach to governance, in contrast to greater emphasis on markets and more fragmented and targeted regulation in the United States.

Box 3.2
Informed Consent Around the World

- The European Commission (EC) provides granular details about what must be included in an informed consent. Such details include "Alternative procedures or treatments that might be advantageous to the participant need to be disclosed" and "Procedures [researchers will take] in case of incidental findings."
- Many African countries do not require informed consent, sometimes due to lack of governance structures and research infrastructure and sometimes due to the desires of an authoritarian regime. Regardless, researchers from elsewhere in the world may use this gap as an opportunity to exploit local research subjects.
- We found difficulty examining informed consent in China, as cases exist where informed consent was not upheld to the standards described in Chapter Two, and a lack of data prevents us from determining whether these cases represent normal practices.

SOURCE: EC, *Ethics for Researchers*, 2018a.

---

[21] Interview 3.

[22] Canada's Panel on Research Ethics hosts a Tri-Council Policy Statement (Canadian Panel on Research Ethics, *Tri-Council Policy Statement: Ethical Conduct for Research Involving Humans*, Ottawa, Canada: Government of Canada, 2014) shared by three research-funding bodies that originated in 1998 and was updated in 2010 and 2014 to address, among other things, multijurisdictional research.

[23] European Commission, "Research and Innovation," webpage, 2018d.

[24] "AI4EU Project," webpage, EU, 2018.

The EC's report *Ethics for Researchers* says, "There is a strong connection between research ethics and human rights."[25] See Box 3.3 for excerpts from the European Charter on Fundamental Rights, which aligns closely with the ethics defined in Chapter Two. The EC identifies three research areas it will not fund: human cloning, research that modifies the genetic heritage of human beings, and research that will create human embryos solely for the purpose of research or stem-cell procurement. Member states within the EU can conduct research permitted by their domestic laws, but any research funded by the EC will adhere to its restrictions, which represent terms agreed across the members.[26] On stem-cell research and human embryos specifically, different EU member countries have different laws, allowing this research to occur in some countries but not others.[27] Meanwhile, in the emerging area of artificial intelligence ethics, the United Kingdom has signaled an interest in becoming a leader in ethical artificial intelligence.[28]

India and China attract attention because of their efforts to do more research and to leverage research as part of their economic development. Priorities and resources can lead to gaps between policy or codes of ethics and practice. Anecdotal and some

---

Box 3.3
The European Charter of Fundamental Rights

The European Union is founded on a common ground of shared values laid out in the European Charter of Fundamental Rights, which contains several principles relevant in the context of research. These principles form the basis of important ethics guidelines but also support the conduct of research.

Article 3 – Right to the integrity of the person

Everyone has the right to respect for his or her physical and mental integrity.

In the fields of medicine and biology, the following must be respected in particular:

- the free and informed consent of the person concerned, according to the procedures laid down by law
- the prohibition of eugenic practices, in particular those aiming at the selection of persons
- the prohibition on making the human body and its parts as such a source of financial gain
- the prohibition of the reproductive cloning of human beings.

Article 7 – Respect for private and family life

Everyone has the right to respect for his or her private and family life, home and communications.

Article 8 – Protection of Personal Data

Everyone has the right to the protection of personal data concerning him or her.

Such data must be processed fairly for specified purposes and on the basis of the consent of the person concerned or some other legitimate basis laid down by law. Everyone has the right of access to data that has been collected concerning him or her, and the right to have it rectified.

Compliance with these rules shall be subject to control by an independent authority.

Article 13 – Freedom of the Arts and Sciences

The arts and scientific research shall be free of constraint. Academic freedom shall be respected.

SOURCE: EC, 2018a, p 9; adapted.

---

[25] EC, 2018a.

[26] EC, 2018a.

[27] EuroStemCell, "Regulation of Stem Cell Research in Europe," webpage, 2018.

[28] Parliament of the United Kingdom, "UK Can Lead the Way on Ethical AI, Says Lords Committee," April 16, 2017.

more-formal evidence suggest that when problems arise in research where protections of human subjects have been inadequate, a backlash can motivate new policy.[29] For example, India's growth of a generic pharmaceuticals industry has raised questions about human subjects protection; enforcement may have improved as a result, but systemic change is difficult.[30]

China has attracted internal and external health tourism on the basis of stem-cell treatments not supported by the kind of testing Western medicine encourages, with publicity appearing to motivate an increase in oversight.[31] Similarly, critical reactions to harvesting organs from prisoners led to curtailment of that practice.[32] China's scientific establishment has connected with the American Association for the Advancement of Science and others for assistance in promoting research integrity.[33] These past events, combined with new commitments to rising stature in the research community, may motivate progress, though attitudes toward research on human embryos are more permissive in China than in the West.[34] That said, recent events in China involving a researcher who claims to have conducted gene editing on embryos that led to live births and outrage in both China and abroad demonstrate both that individual researchers might not follow codes of conduct or laws and that Chinese ethics and expectations have been evolving.

> China's population as a whole does not hold religious, ethical, or other beliefs that would be an obstacle to [embryonic stem cell] research. The Chinese government considers [embryonic stem cell] research to be a strategic, emerging technology and has established several national science and technology programs to support its development. In December 2003 the Ministry of Science and Technology and the Ministry of Health issued *Ethical Guiding Principles on Human Embryonic Stem Cell Research* to codify the ethical principles guiding China's [embryonic stem cell research].[35]

---

[29] In the early 2000s, reports of human subject abuses in China and in India led to improved oversight conditions, according to Interview 4.

[30] "It is futile here to merely repeat the numerous examples which so convincingly demonstrate how these ethical principles have often been violated unless we can unravel the underlying reasons. And the reasons lie in the fact that the 'de jure' principles have to operate in the given political, economic, and social conditions of the real world which dictate the "de facto" ethics of the society." (V. Bajpai, "Rise of Clinical Trials Industry in India: An Analysis," *ISRN Public Health*, Vol. 2013, Article 167059, 2013).

[31] J. Qiu, "Injection of Hope Through China's Stem-Cell Therapies," *Lancet Neurology*, Vol. 7, No. 2, 2008, pp. 122–123.

[32] S. Denyer, "China Used to Harvest Organs from Prisoners. Under Pressure, the Practice Is Finally Ending," *Washington Post*, September 15, 2017.

[33] American Association for the Advancement of Science, "AAAS, China, and Ethics in Science," 2018.

[34] Interview 12.

[35] Interacademy Partnership, *Doing Global Science: A Guide to Responsible Conduct in the Global Research Enterprise*, Washington, D.C.: U.S. National Academies of Science, Engineering, and Medicine, 2016.

Differences in approaches among countries can complicate international collaborations generally, and they can also aggravate the challenge of investigating allegations of misconduct in research.[36]

Ethics "Dumping"

Within the scientific research and ethics communities, a distinction is made between accounting for regional differences and taking advantage of regional differences.[37] Ethics dumping is a kind of context-arbitrage; it is the practice of researchers trained in cultures with rigorous ethical standards traveling to conduct research in countries with lax ethical rules and oversight, not to study a foreign indigenous people but to circumvent the regulations, policies, or processes that exist in their home countries.[38] While such behaviors are considered exploitative and unethical by many, their defenders argue that moving clinical trials to industrializing nations is a response to the conservatism and high data demands of the U.S. Food and Drug Administration and to the statistics required to demonstrate the efficacy of drugs offering incremental improvements over existing alternatives. Defenders also argue that the populations in these countries tend to present fewer potential drug interactions than those in industrialized nations.[39] Although researchers may travel to countries without strong rules or governance on the grounds that research there is less expensive or involves less bureaucracy, the bypassed processes are designed to protect research participants, and circumventing them may exploit local populations.[40] One contested issue is the use of placebos, which has been limited by the Declaration of Helsinki. The issue relates to whether placebos in effect withhold treatment and potentially jeopardize participants' health. The FDA shifted to alternative guidelines developed by a group of European, Japanese, and U.S. regulators along with the pharmaceutical industry that base the standard of care on what people would receive locally, which effectively increases the use of placebos in areas with low standards of care and raises corresponding ethical questions.[41]

---

[36] Interacademy Partnership, 2016.

[37] In this discussion, we intentionally avoid terms like *First World* and *Third World countries* or *developed* and *undeveloped* countries, because those distinctions are misleading. There may exist developed, First World countries with less-regulated ethics and governance than certain less-developed countries.

[38] The term *ethics dumping* was possibly first used by the EC (EC, "Reducing the Risk of Exporting Non Ethical Practices to Third Countries," GARRI-6-2014, request for proposals, December 10, 2013).

[39] K. Weigmann, "The Ethics of Global Clinical Trials in Developing Countries: Participation in Clinical Trials Is Sometimes the Only Way to Access Medical Treatment. What Should Be Done to Avoid Exploitation of Disadvantaged Populations?" *Embo Reports*, Vol. 16, No. 5, 2015, pp. 566–570.

[40] Interview 5.

[41] Weigmann, 2015. The guidelines can be found in International Council for Harmonisation, "ICH Guidelines," undated.

These situations bring into focus some of the language used in the principles discussed in Chapter Two. How should one think about the nature and mix of "beneficence," "duty to society," and "integrity," for example, when individuals with limited financial capacity and limited education find economic as well as medical benefit from participating in clinical trials and the development of relevant infrastructure benefits local communities? Contrasting circumstances in different parts of the world affect the cost-benefit analyses that undergird some research ethics in practice.

One antidote to ethics dumping is expecting the same rules, standards, and practices of researchers that would obtain in their home countries; that is the approach taken by U.S. and European governmental research funders. The EC, for example, requires that researchers who conduct research in other countries adhere to the same standards as they would in Europe, specifically,

> The research needs to comply with all the relevant European legislation, national legislation and with relevant accepted international standards.

- International research projects must be beneficial for all stakeholders, with emphasis on benefits for the research participants and their communities. Special initiatives to support local communities (e.g. provide access to basic health care and the benefits generated by the research), can help achieving this goal.
- If local resources are used, this should be adequately compensated.
- Potentially vulnerable populations need to be able to provide **genuine** informed consent [emphasis in original text]. This requires taking into account potential cultural differences, economic and linguistic barriers and levels of education and illiteracy.
- Although adequate scientific and ethics infrastructure might not be available, the relevant local and independent ethics approvals need to be provided.[42]

Another antidote, which acknowledges cultural differences, is for researchers to work with local communities in framing research—an exercise in community-based participatory research,[43] which takes place within Europe and the United States. An example of such research is the development of the San code of ethics discussed in Chapter Two, which was supported by the three-year EU TRUST project.[44] TRUST aims to create standards for research around the world, and it has produced a Global Code of Conduct for Research in Resource-Poor Settings, which was introduced in

---

[42] EC, 2018a.

[43] Weigmann, 2015.

[44] TRUST Project, 2018.

2017 and is being promoted within the EU.[45] It puts a spotlight on long-term interests over short-term expedience.

International Standards Development

Growing attention to privacy, security, and embedded biases associated with software systems (including but not limited to those associated with AI) has led to renewed attention to ethics in computer, information, and data science, for both related research and professional practice. These fields have a strong international character. Leading academic programs in the United States have long attracted students from other countries, especially from China and India, and the leading professional societies that are headquartered in the United States see themselves as international. Two major professional societies, the Institute for Electrical and Electronics Engineers (IEEE) and the ACM, recently updated and enhanced codes of ethics to address new concerns associated with AI.

Unlike other processes discussed in this chapter, which connect to governments, at least in their origin or authority, professional societies have more of a grassroots, voluntary character.[46] The recent progress of the IEEE Global Initiative on Ethics of Autonomous and Intelligent Systems is illustrative. Seeded by IEEE leadership interest, the work attracted 100 people (growing to 250) addressing ethics-related topics through committees (13 as of this writing)[47] and engaged a set of additional groups that provided language translation and an interface to interested parties in China, Japan, South Korea, Brazil, Thailand, Russia, and Hong Kong.[48] The initial work has been based on consensus or majority agreement. A more formal process would be invoked if the work were to advance to an IEEE policy.

Concluding Observations

Research ethics have been evolving in ways motivated and shaped by phenomena around the world. Whereas the abuses of the Nazis galvanized collaboration across an international community of physicians and medical researchers, problematic practices in individual countries have inspired local improvements in oversight that are believed to be associated with the inculcation of ethical principles. Among industrial-

---

[45] EC, 2018b.

[46] That observation is qualified by the recognition that in some countries (notably China), research and professional activities may be connected to government support and interests.

[47] IEEE, "The IEEE Global Initiative on Ethics of Autonomous and Intelligent Systems," 2017.

[48] IEEE, 2017.

ized nations, reflective and group inquiries have produced seminal documents articulating important principles for both individual countries and groups of countries, albeit with differences on concepts as well as practice.

Professional societies are an important mechanism of self-governance that can be connected to official governance. The modern practice began with physicians who organized the WMA and gave rise to the Declaration of Helsinki, and it continues with the IEEE and others. Professional societies guide the intergenerational transfer of knowledge, including appreciation for ethics for both research and practice, through their role in shaping education and training processes, from the smallest laboratory with a single faculty member engaging students to the academic program that secures discipline-based accreditation.[49] The recent joint development of guidance on ethical conduct of research by an international group of national academies of sciences was an amplification of the role of professional societies.[50] Its goal was to guide young researchers in a world increasingly characterized by team science, cross-disciplinary collaborations, and international collaboration.

EU activities such as the TRUST project demonstrate the influence of funding relevant activities; by contrast, people typically participate in professional societies and standards-setting as volunteers. Work under the auspices of the United Nations on access to benefits from genetic materials is another illustration of how institutional support, especially for multilateral, government-to-government agreements, can foster progress.

Differences across countries reflect cultures and norms, education and awareness, and economics—both the funding for enforcement and the incentives perceived by researchers. Biomedical research has driven consideration of ethics globally because the stakes are so obvious. The rise of nanotechnology has been accompanied by attention to ethical, legal, and social implications,[51] while the rise of AI has added to historic concerns about cybersecurity and digital privacy and spawned attention to ethics in associated research. These phenomena are unfolding globally, if unevenly, and it is possible that balance of research ethics concerns may shift as the ethical aspects of physical, computer, information, and data sciences attract more attention.

Finally, as collaborations between researchers associated with different governments continue and grow, specialized instances of knowledge—which may potentially include ethics—are transferred across international borders. The United States has

---

[49] The American Chemical Society, for example, accredits chemistry educational programs at different levels, addressing both safety and ethics. American Chemical Society, "Standards, Guidelines, and ACS Approval Process," webpage, 2018b; American Chemical Society, "Materials for Ethics Education," webpage, 2018a. The Accreditation Board for Engineering and Technology accredits educational programs at different levels across a variety of levels and engineering disciplines (extending to computer science), addressing ethics (Accreditation Board of Engineering and Technology, homepage, 2018).

[50] Interacademy Partnership, 2016.

[51] National Nanotechnologies Initiative, "Ethical, Legal, and Societal Issues," webpage, 2018.

bilateral agreements, typically relating to life sciences, infectious diseases, and civilian health, with more than 50 countries that provide frameworks for cooperation between government researchers in the United States and other countries.[52] Additional agreements may exist covering other kinds of scientific and technical collaboration.

---

[52] U.S. Department of State, "Science, Technology, and Innovation Partnerships," webpage, 2018.

CHAPTER FOUR

Monitoring and Enforcing Ethics

Merely discussing, debating, and documenting ethics gets society only so far. Codes of ethics should be monitored and enforced to identify, dissuade, and punish unethical behavior. Monitoring and enforcing ethics serves several purposes, including removing incentives for researchers to act unethically and revealing topics where ethics should be revised or updated. This chapter discusses how ethics that are documented are monitored and enforced. (Ethics that are not documented, such as societal values, remain a topic for another paper.) We began this analysis with the same approach to the literature review described in Chapter Two and Appendix A, and we augmented our literature review with interviews and with a review of several countries' laws.[1]

It should be noted at the outset that ethics are not laws, although they have been discussed recently in the context of "soft law," which combines different kinds of guidelines and nonlegal codes. The differences between hard law and soft law with respect to ethics are shown in Table 4.1, where hard law refers to national laws, and soft law refers to all other mechanisms for enforcing ethical behavior.[2] Because codes of ethics are a form of soft law, enforcement varies in stringency from a scientist being expelled from a professional society, which may or may not affect his or her career going forward, to a physician losing his or her medical license either permanently or temporarily, a career-altering step that can involve national or local legal action. We evaluated these differences along a spectrum of how harsh the punishments could be to the researcher, and we provide examples in Table 4.1.

People interviewed for this project emphasized the importance of both bottom-up inculcation of concern about ethics and a credible threat of sanctions, such as from some kind of regulator. The variations among both disciplines and cultures militate against the notion of a universal code, and both promotion and enforcement of ethics are accordingly decentralized. When there are alternatives, as has been noted by schol-

---

[1] This document was helpful to us during our research of foreign laws: HHS, 2018.

[2] Interview 8.

Table 4.1
Codified Ethics in Hard Law and Soft Law

		Codification Method	Mechanism for Enforcement
Harsh repercussions for the researcher	Hard law	National laws, state laws, and local laws	Investigation and prosecution can lead to prison, probation, fines, or other punishments.
↓	Soft law	International treaties and ratified agreements (e.g., Biological Weapons Convention)	May result in the writing of national laws. Severe cases of breaches of such treaties without local or national enforcement may lead to prosecutions in the International Criminal Court or sanctions by the United Nations Security Council.[a]
		Rules of certifying associations (e.g., AMA Code of Medical Ethics)	Disciplinary action against members can lead to permanent or temporary revocation of ability to conduct research.
		International statements and declarations (e.g., Declaration of Helsinki)	May result in the writing of national laws.
		Employer codes of conduct (e.g., RAND Corporation's Institutional Principles and Code of Ethical Conduct)	Disciplinary action against employees can lead to loss of employment.
Weak repercussions for the researcher		Rules of member associations and societies (e.g., IEEE Code of Ethics)	Disciplinary action against members can lead to ouster from the group.

SOURCE: RAND analysis and Interview 8.

[a] International agreements vary in their placement on this spectrum. Even where there is a body with responsibility for the implementation of agreements, monitoring is imperfect at best. Cases regarding scientific research ethics that could be elevated to the International Criminal Court would fall under its jurisdiction for prosecuting crimes against humanity, and we know of no such cases that the court has undertaken.

ars concerned with soft law, not knowing how to pick and choose can undermine compliance.[3]

Ethics begin to have influence through processes of education and training. As noted in Chapter Three, professional societies that oversee the accreditation of academic programs in science and engineering include ethics as part of their curricula. The training of researchers goes beyond the classroom or textbook to include the experience in the laboratory, as part of multigenerational teams blending experienced and credentialed faculty or other research leaders with graduate and undergraduate (and sometimes high school) students.

---

[3]   Interview 8.

Mentorship is a key part of this system. For many, the active learning by doing involved in laboratory or other hands-on research has the strongest influence on their careers.[4] Early-career laboratory work is both an engine of biomedical research, in particular,[5] and formative for the individuals involved. Practices from safety to ethics are inculcated by watching and emulating the leader—whose practices may or may not be exemplary. Lab-based teams are also an important vehicle for international knowledge transfer: Academic science and engineering in the United States involves students from many countries, and students who return to native countries bring back what they learn about theory, empirical practice, safety, and ethics.

Researchers reinforce their professional identity through membership in professional societies, which, as documented throughout this report, draw upon their memberships to develop and disseminate codes of ethics. That said, professional societies are voluntary enterprises with no real enforcement capacity—the termination of a society membership is not the most onerous sanction.

Research integrity is an arena that involves ethics beyond the treatment of research participants. It can involve fabrication or falsification of data, plagiarism, appropriation of credit, or abusive supervision of junior researchers.[6] These ethical breaches compromise research quality and undermine trust in the research community, as documented by multiple committees under the aegis of the National Academies of Sciences, Engineering, and Medicine.[7]

Institutions that house research, such as national laboratories, companies, independent laboratories, and universities, have oversight of the conduct of research and the responsibility for adherence to ethics, along with safety, intellectual property, intolerance of harassment, and various other aspects of behavior. These institutions have designated people and committees (e.g., IRBs for human subjects protection and separate bodies focused on research integrity) that oversee the responsible and ethical conduct of research. Oversight is intrinsically reactive; problems need to be detected, investigated, and addressed, each of which involves its own shortcomings. In some instances, training, monitoring, and reporting is connected to funding, or there may be regulatory requirements (e.g., for workplace safety).

---

[4]   Interviews 6 and 11.

[5]   "The great majority of biomedical research is conducted by aspiring trainees: by graduate students and postdoctoral fellows." (B. Alberts, M. W. Kirschner, S. Tilghman, and H. Varmus, "Rescuing US Biomedical Research from its Systemic Flaws," *Proceedings of the National Academy of Sciences of the United States of America*, Vol. 111, No. 16, 2014, pp. 5773–5777.

[6]   Interviews 11 and 14.

[7]   C. K. Gunsalus, A. R. Marcus, and I. Oransky, "Institutional Research Misconduct Reports Need More Credibility," *JAMA—Journal of the American Medical Association*, Vol. 319, No. 13, 2018, pp. 1315–1316; National Academies of Sciences, Engineering, and Medicine, *Fostering Integrity in Research*, Washington, D.C.: The National Academies Press, 2017.

In many if not most instances, ethics (as opposed to safety) training is associated with adherence to human subjects protection (which itself may be linked to funding). The quality of those processes varies, and the rise of a compliance mentality with growing need to demonstrate compliance may be counterproductive, as observed by a group of leading U.S. life scientists: "[E]xpanding regulatory requirements and government reporting on issues such as animal welfare, radiation safety, and human subjects protection . . . are important aspects of running a safe and ethically grounded laboratory . . . [but] are taking up an ever-increasing fraction of the day."[8]

Journals that publish research can play a small role in enforcing ethics.[9] For example, where human subjects or animal protections are expected, reviewers and editors of articles look for evidence of compliance with associated processes. That said, journals are removed from the conduct of the research and depend on the institutions where research takes place for true monitoring and enforcement. When problems have been discovered after publication, articles are retracted,[10] but especially if the quality of the outlet is not a concern, essentially anything can be published somewhere, as is evident on the web more generally. Short of a problem warranting retraction, concerns about ethics do not tend to be published within the research publication itself.[11]

The challenge of achieving ethical research is systemic, and problems can arise in any or all components. As a recent National Academies of Sciences, Engineering, and Medicine report observed,

> Integrity in research means that the organizations in which research is conducted encourage those involved to exemplify these values in every step of the research process: planning, proposing, performing, and reporting their work; reviewing proposals and work by others; training the next generation of researchers; and maintaining effective stewardship of the scholarly record. . . . [R]esearch institutions may—or may not—create and maintain research environments that support integrity, including the policies and capabilities needed to respond responsibly to allegations of research misconduct. Science, engineering, technology, and medical journal and book publishers may provide high levels of rigor in review of manuscripts, or they may put pressure on prospective authors to add citations to manuscripts to improve a journal's score on a bibliometric indicator. Fields and disciplines may take on as a community the task of defining and upholding necessary

---

8   Alberts et al., 2014.

9   Interview 12.

10   The case of an article in a prestigious journal suggesting a link between vaccines and autism, which influenced an enduring antivaccination movement despite being retracted because of its inadequacies, demonstrates the challenge of truly correcting misunderstandings arising from inappropriate publication and the limited capacities of even high-quality journals. See, for example, D. W. Hackett, "16 Year Old 'Vaccines Cause Autism' Paper Withdrawn, Finally," 2018; J. Belluz, "20 Years Ago, Research Fraud Catalyzed the Anti-Vaccination Movement. Let's Not Repeat History," Vox.com, 2018.

11   Interview 12.

standards in areas such as data sharing, or they may fail to do so and, in effect, tolerate detrimental research practices.[12]

The ultimate challenge for fidelity to a code of ethics is incentives. How can researchers be motivated to do the right thing consistently? Do perverse incentives exist? How do circumstances shape the choices a researcher makes, including potential shortcuts or compromises? What kinds of cognitive biases are at play within research teams, research institutions, and research oversight functions? In short, how can commitment to ethics be internalized?

People who study the changing nature of the research enterprise have expressed growing concerns about the pernicious effects of current incentives. Decisionmaking has shifted from how to do things the right way to how to get things done.[13] Beyond the financial rewards typically associated with conflicts of interest, institutions responsible for oversight increasingly attend to reputation and the benefits associated with engaging high-stature researchers, which may lead to the proverbial blind surveillance eye.[14] More broadly, constraints on funding have produced high levels of competition for grants, jobs, and promotions, altering the climate in the research laboratory and generating pressure "to rush into print, cut corners, exaggerate . . . findings, and overstate the significance of . . . work."[15] The growing investment in research of countries such as China add to the competitive atmosphere.[16]

A counter to the troubling circumstances in institutionalized research outlined above comes from the more freewheeling environment of DIY Bio (discussed further in Chapter Five) combined with more-formal biotechnology research, which presents risks of problems from both accidents and malice. Perhaps because those risks are so compelling, U.S. government personnel have had success in raising awareness and promoting training and self-monitoring in research communities. The resulting sense of responsibility involves an active embrace of ethical principles, even if the discourse is different from that of Chapter Two.[17] The attitude fostered is one of "not on my watch," as opposed to "do no harm."

---

[12] National Academies of Sciences, Engineering, and Medicine, 2017.

[13] Interview 12; C. D. Gunsalus, A. R. Marcus, and I. Oransky, "Institutional Research Misconduct Reports Need More Credibility," *JAMA—Journal of the American Medical Association*, Vol. 319, No. 13, 2018, pp. 1315–1316; C. K. Gunsalus and A. Robinson, "Nine Pitfalls of Research Misconduct," *Nature*, Vol. 557, 2018, pp. 297–299.

[14] Gunsalus et al., 2018, citing an Institute of Medicine Report.

[15] Alberts et al., 2014.

[16] Interview 12.

[17] Interview 13.

CHAPTER FIVE

Emerging Ethics Topics

Our research indicates that ethics are created, change, and evolve due to significant historic events that create a reckoning (e.g., the Nuremberg Code), due to ethical lapses that lead researchers to create new safeguards (e.g., the Tuskegee Study), due to scientific advancements that lead to new fields of research (e.g., the emergence of experimental psychology), or in response to changes in cultural values and behavioral norms that evolve over time (e.g., perceptions of privacy and confidentiality). This chapter focuses on these last two topics—new scientific advancements and changes in cultural values—and attempts to anticipate where changes to ethics may be needed in the near future.

Bystander Risk

Concerns about and protections for human subjects focus on people who participate in research, how they consent, how they are treated, and so on. New concerns are now beginning to arise about impacts of research on *other* people, so-called *bystanders*, who were not the research participants and who did not consent to participate. These new concerns point to variations or ambiguities in how researchers interpret informed consent and beneficence.

People who choose not to participate in genetic testing and genetic research have worried about the repercussions when other members of their family bloodline agree to participate. In the United States, the Genetic Information Nondiscrimination Act (GINA) of 2008 prevents genetic information from being used against a person in health insurance and employment decisions, but it does not protect against the use of such information for other types of insurance or for other uses.[1] Laws like GINA do not exist in all countries, and even in the United States, genetic material collected out-

---

[1]   National Human Genome Research Institute, "The Genetic Information Nondiscrimination Act of 2008," April 17, 2017.

side of research purposes has been used as evidence for criminal arrests, indicating the limitations of such laws regarding privacy.[2]

In medical research, the potential for secondhand exposure to a contagious disease from someone who did consent to participate in research is another topic for concern. Several researchers recently served on an expert panel to determine whether healthy volunteers could ethically be deliberately infected with Zika virus. They described their experience specifically through their concerns for bystander risk: What happens if a research participant who consents to being exposed to Zika infects a second person through sexual contact? What happens if that exposed person becomes pregnant, exposing a fetus to Zika? These secondary and tertiary exposures (the sexual partner and fetus, respectively) put bystanders who lacked the opportunity to consent to participating in the initial Zika trial at risk.[3] Such bystanders would gain none of the benefits from the research, risk taking on all of the potential consequences, and lack an opportunity to consent to participation.

In other instances, the privacy and potentially the physical well-being of individuals may be compromised by autonomous systems in their environs. Drones and automated vehicles operate by using cameras, other kinds of imaging systems, and even microphones as part of their systems for perceiving their environment. How, by whom, when, and where sensory information is collected, stored, and used has implications for people whose images and locations are captured incidentally. These people are true bystanders, in the usual sense of the term. Their situation illustrates the limitations of informed consent—it is impossible to predict who will be in the path of an experimental or commercial unmanned system (outside of a tightly controlled campus context).

A final example comes from social media. Computer science and other research has shown that it is possible to learn about people beyond what they disclose voluntarily by analyzing whom they connect to.[4] People who consent to share information only with their "friends" or "followers" do not understand the broader consequences of such sharing. There is a larger set of questions that comes from wholesale consideration of social media (see discussion of Big Data below), which may be facilitated by a service provider offering access to researchers (e.g., through an application program interface [API]) or developed directly through some kind of web scraping. As this report is being written, criticisms of an absence of control over the use of user information and the launch of the EU's General Data Protection Regulation (GDPR) may result in new

---

[2]   M. Berman, J. Jouvenal, and A. Selk, "Authorities Used DNA, Genealogy Website to Track Down 'Golden State Killer' Suspect Decades After Crimes," *Washington Post*, April 26, 2018; E. Shapiro, "'I Honestly Never Thought They Would Find Him': DNA Test, Genetic Genealogy Lead to Arrest in Woman's 2001 Killing," ABC News, November 6, 2018.

[3]   S. K. Shah, J. Kimmelman, A. D. Lyerly, H. F. Lynch, F. G. Miller, R. Palacios, C. A. Pardo, and C. Zorrilla, "Bystander Risk, Social Value, and Ethics of Human Research," *Science*, Vol. 360, No. 6385, 2018, pp. 158–159.

[4]   See, for example, I. Kloumann and J. Kleinberg, "Community Membership Identification from Small Seed Sets," Cornell University, 2014; C. Y. Johnson, "Project 'Gaydar,'" *Boston Globe*, 2009.

practical restrictions and test what kind of informed consent might be possible.[5] In the meantime, people around the world are considering the public-space metaphor that has been used, and IRBs are considering how to keep up with ideas for research using social media.

Big Data

Big data can raise questions for research ethics. The concerns tend to revolve around the potential to compromise privacy. Part of what makes big data so large is the combined use of multiple sources of data. In particular, even if an initial data set has been de-identified, it is possible to re-identify subjects through the use of new information—meaning that traditional approaches to and expectations for de-identification can be vitiated by big-data analytics. Another challenge from big data is that they can be accumulated by bringing together data collected for a variety of reasons and using the data in new ways. People have noted that where direct discrimination is avoided, discrimination can occur indirectly by relying on data associated with people having certain attributes (although this can be done for helpful and harmful effect)—looking, for example, at shopping patterns and what they imply for demographics or for creditworthiness.[6] As such an example shows, discrimination can affect groups as well as individuals.

Increasingly, data will be used for reasons other than what motivated their original collection. Whereas data collected about people through traditional research may be associated with informed consent and controls on secondary use, big data may involve unanticipated secondary uses of a wide variety of data, including data (such as locational metadata) that may not seem privacy-sensitive.

> In our digital society, we are followed by data clouds composed of the trace elements of daily life—credit card transactions, medical test results, closed-circuit television (CCTV) images and video, smart phone apps, etc.—collected under mandatory terms of service rather than responsible research design overseen by university compliance officers. . . . [T]hese informal big data sources are gathered

---

[5]   GDPR is set of data privacy regulations that were implemented across the EU in 2018. A new article discussed the potential of GDPR to limit the uses of citizen science for health-related research, seeking to balance the benefits and burdens of GDPR (A. Berti Suman, and R. Pierce, "Challenges for Citizen Science and the EU Open Science Agenda Under the GDPR," *European Data Protection Law Review*, Vol. 4, No. 3, 2018.

[6]   These and other potential concerns were flagged by the 2014 discussions of Big Data and Privacy by the White House and the President's Council of Advisors on Science and Technology (PCAST) (Executive Office of the President, President's Council of Advisors on Science and Technology, *Report to the President: Big Data and Privacy: A Technological Perspective*," May 2014).

by agents other than the researcher—private software companies, state agencies, and telecommunications firms.[7]

This variety of data sources is one reason the role of IRBs in big data research is limited.[8] It is also one of the ways that big data challenge traditional fair information practice principles (see Box 2.2). Another is the potential for big data analytics to generate outputs—to create new data that were not provided by the research subject, but rather were created about him or her. One such example is financial credit scores or other such scoring and rating mechanisms used by financial institutions or insurance companies. These ratings may be privacy-sensitive but calculated from inputs that were not privacy-protected.

A factor in the rise of big data is the rise of open data—in particular, data collected using government resources and made available for broad use. Because of the range of data sets that can be used ad hoc or opportunistically, and because an increasing amount of big data are collected from the environment (the growing Internet of Things that proliferates cameras and microphones), some have suggested a shift from ex ante informed consent to an approach that focuses on harms that might be generated.[9] More generally, given the range of research contexts for big data, a group of researchers, including ethicists and funded by the National Science Foundation, has advocated that each research community should debate and develop a code of conduct relating to its use of big data.[10]

In the biomedical arena, very large cohorts are being assembled that will generate their own big data. For example, in support of precision medicine, the U.S. National Institutes of Health has been developing a million-person cohort,[11] offering trust and privacy principles and a public protocol as part of an elaborate process to build in ethics.[12] Meanwhile, questions have arisen about how potentially very large collections of data, such as data about people who use consumer-oriented genetic sequencing services, are handled and protected. Consumers using these services bought a service with

---

[7]   M. Zook, S. Barocas, D. Boyd, K. Crawford, E. Keller, S. P. Gangadharan, A. Goodman, R. Hollander, B. A. Koenig, J. Metcalf, A. Narayanan, A. Nelson, and F. Pasquale, "Ten Simple Rules for Responsible Big Data Research," *Plos Computational Biology*, Vol. 13, No. 3, March 30, 2017, pp. 1–10.

[8]   Zook et al., 2017.

[9]   Executive Office of the President, PCAST, 2014.

[10]   Zook et al., 2017. The National Science Foundation funded the Council for Big Data, Ethics and Society, which has generated case studies and other guidance materials (Council for Big Data, Ethics and Society, homepage, undated).

[11]   National Institutes of Health, "About," webpage, 2018a.

[12]   National Institutes of Health, "All of Us Program Protocol," webpage, 2018b.

no expectation that they were contributing to an unknown body of research and with no opportunity to provide an informed consent or to refuse consent.[13]

New paradigms for addressing concerns that can be associated with large groups of people—extending to whole communities or some concept of society at large—may be needed. For example, a researcher at the Massachusetts Institute of Technology, Iyad Rahwan, has suggested that just as conventional computer-based systems may be designed for a human role in their control—a human in the loop—it may be possible to design to have society in the loop.[14] What "society" means, who decides or interprets it, and how much the meaning might vary across communities or cultures would be among the questions to address—having society in the loop implies balancing the interests of different stakeholders. Today's debates about platform technologies demonstrate different outlooks on the social aspects of these systems in different countries and regions. Although some of those differences are rooted in culture, others reflect political concerns about global enterprises and where they are headquartered that complicate the analysis. By contrast, the discussions of society that motivated the Uppsala Code discussed in Chapter Two had a different character and focused on the physical survival and well-being of humankind.[15]

Open Science

Open science is a term that refers to exposure and disclosure of key aspects of research to facilitate both access to the results and understanding of how those results were achieved. It has been evolving along a variety of paths. The U.S. federal government[16] and the EC[17] have both advocated for open science.[18] Our sponsor, the Intelligence Advanced Research Projects Activity (IARPA), has expressed its commitment to openness associated with its own work.[19] Ethical issues can arise when sharing processes

---

[13] As of December 2018, customers could opt out of having their data shared with public research, but they could not opt out of internal company research (CitiGen, "What Happens to Your Genetic Data When You Take a Commercial DNA Ancestry Test?" 2017.

[14] I. Rahwan, "Society-in-the-Loop: Programming the Algorithmic Social Contract," *Ethics and Information Technology*, Vol. 20, No. 1, 2018, pp. 5–14.

[15] Related discussions today often center on environmental concerns, such as the effects of climate change.

[16] Office of Science and Technology Policy, "Memorandum for the Heads of Executive Departments and Agencies," 2013.

[17] EC, "Open Science," webpage, 2018c.

[18] The EC commissioned RAND Europe to develop a tool for monitoring EU open science: E. Smith, Salil Gunashekar, Sarah Parks, Catherine A. Lichten, Louise Lepetit, Molly Morgan Jones, Catriona Manville, and Calum MacLure, "Monitoring Open Science Trends in Europe," Santa Monica, Calif.: RAND Corporation, TL-252-EC, 2017.

[19] Office of the Director of National Intelligence, "Public Access to IARPA Research," webpage, 2018b.

are manipulated to protect perceived advantages from withholding data, code, or even results.

Open Data

Open data refers to sharing data associated with research (including open and shared code). It typically is associated with open science because easy availability of data facilitates the review and understanding of research and the conduct of new research, including research intended to reproduce results. The authors of an article titled "Data Sharing by Scientists: Practices and Perceptions" argue that "[d]ata needs to be stored and organized in a way that will allow researchers to access, share, and analyze the material. . . . The development of data standards may provide a foundation for cross-community collaboration in format and ontology development, making it much easier for laboratories to manage, integrate, and analyze data."[20]

Open sharing of data or code has been resisted in some contexts where researchers believe that their rewards will derive from their control over data and/or code; research funder mandates for data and code sharing are intended to combat that attitude, but their impact tends to be on publicly funded research. At least some privately funded research is not disclosed at all, let alone associated data and methods, because of its proprietary value (although some companies have begun to make internally developed tools[21] available for broad research use—an act that lowers research costs for users while linking them to particular infrastructure). In some instances, privacy and other concerns about the sensitivity of data have inhibited sharing, consistent with ethical principles discussed elsewhere in this report; mechanisms for masking sensitive aspects of data are the subject of their own research, given the power of big-data analytics to unmask, as noted above.

Open Publication

Open science is often associated with open publication. Traditional research journals have asserted strong intellectual property rights over the content they publish, with one possible unintended consequence being limitations on who can afford to access research articles. One pioneering response to the combined cost and delay associated with traditional journal publication emerged from physics and has begun to expand to other disciplines: the informal development of an archive for research made available prior to its formal publication.[22]

Open publication has given rise to two concerns. First, because these new approaches to open science are low- or no-cost to those who want to read associated

---

[20]   C. Tenopir, S. Allard, K. Douglass, A. U. Aydinoglu, L. Wu, E. Read, M. Manoff, and M. Frame, "Data Sharing by Scientists: Practices and Perceptions," *PLoS ONE*, Vol. 6, No. 6, 2011.

[21]   See, for example, Google's tools for AI (Google, "Google AI," tools, 2018).

[22]   Cornell University, ArXiv, 2018.

research, researchers typically pay a fee to offset the costs of processing their articles. That circumstance has raised questions about the potential for conflicts of interest[23] (what kinds of science are available where journals publish research only for those who pay?) or excessive fees (will journals be incentivized to publish low-quality or unethical work to earn revenue?).[24] Provisions for prepublication review, which has been a hallmark of traditional journal publication and an expectation of members of the research community (that is, people expect to be reviewed and to provide reviews), vary widely in open-access publications. Although the ethos of open science can involve a shift from a closed set of reviewers to an expectation that the community that reads the research is effectively reviewing and evaluating it, true peer review—review by people who know something about the subject or methods of the research—may occur but with more variability in its operation. Although peer review itself has been criticized for how it works in practice, it is part of the gatekeeping that has put a floor under traditional journals, while sales of access to journals, in turn, are a key part of the business model of the professional societies on which researchers depend for community-building, development of community ethics, accreditation of education and training, and so on.[25] In short, although open science is a mechanism for expanding participation in and consumption of research, its processes remain a work in progress.[26]

Citizen Science

Citizen science, DIY Bio, maker spaces, and other activities and facilities that engage noncredentialled individuals in scientific or engineering processes have been growing throughout this century. Citizen science has been linked directly to open science, especially in the EU.[27] Although it tends to be discussed as new, the history of science was shaped by motivated, curious individuals, and the institutionalization of science was largely a product of the 20th century. Because of that formalization, credentialed scientists sometimes look askance at amateurs engaging in research, and while those amateurs might not comport with the principle of professional competence discussed in Chapter Two, that does not mean that they are beyond the pale of codes of ethics overall. Citizen science provides opportunities for the community to participate in sci-

---

[23] PLOS has an initiative to address unequal capacity to pay around the world: Public Library of Science, "PLOS Global Participation Initiative," FAQ, 2018.

[24] D. N. Salem and M. M. Boumil, "Conflict of Interest in Open-Access Publishing," *New England Journal of Medicine*, Vol. 369, No. 5, 2013, p. 491.

[25] Interview 12.

[26] K. Worlock. "Access to the Literature: The Debate Continues," *Nature*, 2004.

[27] E. Smith, Sarah Parks, Salil Gunashekar, Catherine A. Lichten, Anna Knack, and Catriona Manville, "Open Science: The Citizen's Role and Contribution to Research," Santa Monica, Calif.: RAND Corporation, TL-252-EC, 2017.

ence, thereby providing the chance to level the field of discrimination by allowing the research population to become researchers.[28]

Digital technologies and a variety of web-based platforms have stimulated citizen science, enabling projects shaped by individuals seeking to address environmental and other problems in their communities (in addition to connecting volunteers to researchers in need of extra help with data collection or analysis).[29] These platforms assist with training, quality assurance for data-gathering and analysis, and development of a sense of community. They could be vehicles for at least informing their users about relevant ethics.

DIY Bio benefits from the digitization of processes that makes equipment less expensive, more integrated, or both, facilitating use by nonexperts.[30] Facilities such as community laboratories and maker spaces (facilities where anyone can go to use specialized equipment such as machine tools, 3D printers, or other gear useful in prototype fabrication) may have some kind of training, mentorship, and governance and oversight structures that articulate and attempt to enforce ethical principles. These structures differ from IRBs, although they can be gatekeepers.[31] The potential for people to do DIY Bio in their own homes or other private sites outside of community labs implies a step away from any oversight or gatekeeping. DIY Bio has raised concerns about unintended consequences, such as the accidental release into the environment of a genetically modified organism that either affects the ecosystem or is pathogenic or both, or some kind of safety-related accident.[32]

Finally, the democratization of computer science as a research discipline has seen debates over development and use of software, where some citizen science researchers ascribe to codes of ethics that are promulgated by computer science professional societies, while others do not. The potential for so many people to develop new information technology has given rise to what Adam Thierer has called "permissionless innovation,"[33] a label that raises questions about the ability of different kinds of organizations or individuals to contribute to society and, correspondingly, the role of different kinds of rules, governance, and ethics. Although the proliferation of new kinds of information technology occurs in the world of practice, the core ideas come from research done both formally and informally.

---

[28] R. Chari, L. J. Matthews, M. S. Blumenthal, A. F. Edelman, and T. Jones, "The Promise of Community Citizen Science," Santa Monica, Calif.: RAND Corporation, PE-256, 2017.

[29] Chari et al., 2017.

[30] Gene sequencing is now being done in some middle- and high-school science classes.

[31] IRBs are typically associated with institutions hosting research but can also exist independently.

[32] D. T. Holloway, "Regulating Amateurs," *The Scientist*, 2013. Similar concerns are raised by hacker communities—people who enjoy making and breaking systems and may be teaching themselves and exploring in a hands-on way, with a risk of unintended consequences even without malicious intent.

[33] Mercatus Center, "Permissionless Innovation," webpage, 2018.

CHAPTER SIX

Conclusions

When researchers embark on new research paths, our analysis shows that key ethical principles exist that will remain relevant in uncharted scientific territories. Those principles fall into three broad categories: ethical scientific inquiry; ethical conduct and behavior of researchers; and ethical treatment of research participants. Culturally appropriate formulation of research, with input from affected communities, cuts across all three. Together, ethical principles are intended to foster responsible and reliable research, while avoiding exploitation either of people—who do not understand the situation, who lack resources and are more willing to do things that individuals who can afford other options would be unlikely to do, or who have no knowledge that their information may be used for research—or of the environment.

Questions about the ethics of a given choice or how to apply these ethical principles in new research situations can be answered with help from knowledge that has been accumulating about research ethics and from a variety of institutional resources. Researchers can lean on an array of key pillars: education and training; professional societies and communities that promulgate and advocate for codes of ethics; and governance mechanisms that range from institutional oversight (e.g., focused committees) to formal laws and regulations.

Our research also has shown that the choices researchers make in framing, undertaking, and publishing their work are shaped by a complex mix of incentives that can either foster or militate against ethical conduct. The same institutions that can enforce sanctions against inappropriate conduct also present incentives for making one's research look better than it may actually be. As a result, the existence of ethical codes is necessary but insufficient. Ethical research also requires internalizing a commitment to it, aided by training and education on codes and appropriate research methods, mentoring and workplace cultures that foster ethics, transparency about how the research was conducted, and forums (in person and in writing, local and international) where researchers can share their experiences and the challenges they face.

Finally, the multifaceted globalization of research presents new challenges and opportunities for what can be learned and created and for pursuing new horizons ethically. The history of the ethical principles we highlight is that reactions to exploitative situations involving or flowing from research move from outrage to statements of principle to international agreements. Cultural differences remain, affecting the valuation of individuals relative to communities and discussions of whether the ends justify the means. At a minimum, understanding those differences and considering ways to address them are important for realizing the potential from new science and engineering without serious ethical compromises and costs to society.

Methodology

The majority of our analysis is based on reviewing secondary sources (mostly journal articles) and consulting experts. We started by examining literature, and where we found gaps—questions the literature did not answer—we sought additional commentary and documents and conducted interviews with experts. A detailed description of our methodology for the literature review and interviews is described in this appendix.

Literature Review

We sought out written material in stages, beginning formally and then engaging a broader body of scholarly and gray literature. The initial appraisal of literature outlined below was a vehicle for understanding how ethics are discussed across a broad set of fields. It also served as a basis for subsequent collection of information and its analysis.

At the outset of the study, we searched Scopus and Web of Science for the most-cited articles about ethics and research or codes of conduct and research.[1] All of the articles reviewed in this first phase have been cited more than 300 times by other authors; the most-cited article had been cited over 5,700 times at the time of our data collection. Next, we examined the top 200 articles from our search to determine their relevance for our study.[2] This step reduced our list to 103 articles, which were distributed across disciplines as shown in Box A.1. Using that list of scientific disciplines, we searched for their codes of ethics or conduct. Those codes were treated as authoritative decision documents that establish rules for ethics, ethical conduct, and ethical research, while the initial set of journal articles was treated as research, analysis, and

---

[1] We merged the results to remove duplicates, then we sorted the resulting list from most citations to least. In cases where both databases included the same publication with different numbers of citations, we kept the larger of the two numbers.

[2] Our search terms ("ethic," "code of conduct," and "research") resulted in articles where research ethics were neither a central topic nor discussed in any particular depth. For example, among the articles in our list were a neurology article on cognitive impairment, a medical journal article about cost-effectiveness in health care, and a business journal article about online marketing research. Articles that were about research were not necessarily about ethics, and articles about ethics were not necessarily about research.

Box A.1
List of Scientific Disciplines Examined

Mathematics and statistics

Physical sciences and astronomy

Computer science and information science

Engineering

Business management

Ecology and environmental science

General sciences

Social science

Psychology and psychiatry

Medicine, veterinary, and biology

NOTE: Some disciplines—notably medicine, veterinary medicine, and biology—were aggregated because the distinctions between them were too often blurred. "General sciences" includes cross-discipline sciences and sciences that did not easily fit within other fields.

commentary. This methodology yielded 147 documents (103 journal articles and 44 codes of conduct), divided by year and discipline as shown in Figure A.1

We organized the disciplines in Table A.2 based on the degree to which they interact with the human body. Disciplines at the top of this table focus primarily on inorganic topics, whereas disciplines at the bottom of the table involve implants, drugs, and other interventions with humans and animals. In the middle of this spectrum, research disciplines such as business management, ecology, sociology, and psychology may frequently rely on observations of or interactions with human behavior. We expected this ordering to correspond to interest in human subjects (and animal) protection, which looms large in research ethics literature. Medicine (as an aggregate including veterinary medicine and nonenvironmental biology) was the most heavily represented discipline in our initial collection (see Table A.2). This set of disciplines is

Figure A.1
Documents Reviewed by Publication Year

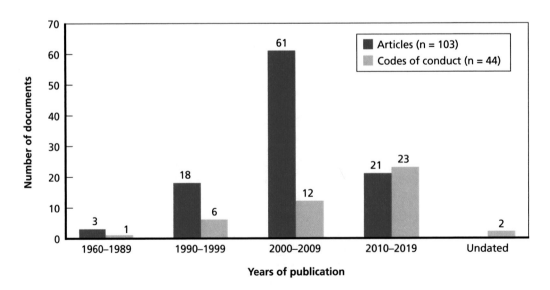

Table A.2
Documents Reviewed, by Discipline

Research Discipline	Articles	Codes of Conduct	Total
Mathematics and statistics	0	3	3
Physical sciences and astronomy	0	4	4
Computer science and information science	6	1	7
Engineering	7	3	10
Business management	6	2	8
Ecology and environmental science	1	2	3
General sciences	15	2	17
Social science	7	7	14
Psychology and psychiatry	1	2	3
Medicine, veterinary, and biology	60	18	78
Total	**103**	**44**	**147**

more likely than other research disciplines to use human participants and animal subjects in clinical trials, resulting in a significant number of articles on the ethical treatment of such subjects. Some of that literature reflects the established nature of the field of bioethics, which has become a model for ethics associated with other disciplines (such as environmental science).

The team generated a codebook, provided in Appendix B, to code and analyze excerpts from the literature and codes of conduct.[3] This codebook assisted in our analysis, though our final list of ethical principles in Chapter Two varies from those listed in the codebook. For example, while bribery was in our codebook, we found that bribery was not discussed in the literature or codes of conduct, possibly because this topic is covered under discussions of conflicts of interest (regarding preventing financial conflicts and disclosing research funding sources) and beneficence (regarding research participants gaining meaningful benefit from the research, which would disqualify bribes for participation). Where there was a concept of possible relevance that didn't fit

---

[3] In 2017, a team of graduate students at Syracuse University's Maxwell School of Citizenship and Public Affairs researched ethical codes of conduct to identify common ethical elements across codes. The students provided their results to the RAND Corporation, and this became the initial framework for our team's codebook, which is provided in Appendix B. We added to the Maxwell students' framework as new topics arose; therefore, the list of topics in the codebook does not exactly match the list of topics in Table 2.1. The student team consisted of Kashaf Ud Duja Ali, Eni Maho, Earl W. Shank, and Derrick J. Taylor, working under the supervision of Professor Renée de Nevers.

our established principles, we coded it "other" for further analysis. Table A.3 lists the number of documents (out of 147 total) that were coded with each topic.

This coding exercise illuminated gaps in our initial literature collection. At times, topics were not adequately addressed in our literature, historical context was lacking, regional and cultural differences from around the world were not discussed, or new emerging scientific and ethical developments were not identified through our methodology. For these reasons, we used additional searches to supplement understanding, and the additional literature we found is cited throughout the report. This additional literature includes two categories: additional journal articles and gray literature. Our gray literature sources include reports and published documents from governments and multinational organizations (including the UN and EC); public laws, treaties, articles from news and business media, key websites, and agreements; and reports published by societies, associations, and other institutions that are not included in peer-reviewed journals.

The documents reviewed in this project are included in three places: Appendix E, which includes our coded-literature bibliography of the 91 journal articles we reviewed; Appendix F, which includes the 40 codes of conduct we collected and reviewed; and the References section, which contains any documents reviewed that do not fall into the previous two categories.

Table A.3
Literature Review Results

Code	Total Number of Documents
Adherence to code	37
Beneficence and nonmaleficence	63
Conflict of interest	33
Data management	33
Duty to society	36
Informed consent	42
Integrity	51
Nondiscrimination	27
Nonexploitation	22
Other topics (not defined)	17
Privacy and confidentiality	39
Professional competence	61
Professional discipline	28

Interviews

Once our initial literature review and coding analysis were completed, we undertook a series of interviews. We selected experts on research processes, ethics associated with research, and a broad set of concerns associated with the research ecosystem, some of which impinge on ethical choices by researchers. In addition to these factors, we sought interview participants who could describe these factors globally or for regions outside the United States. Our interview participants reside in the United States, Europe, and China; interviewees who are based in the United States had broad and sometimes detailed knowledge of ethics issues around the world. These interviews were semistructured, allowing for us to adapt to the particular expertise and interests of the participants. The information obtained was valuable for providing nuance and interpretation of trends and circumstances in the research community, the research-funding community, and professional societies. Beyond giving perspective on the initial literature review, the people with whom we spoke guided us to additional sources that can be found in the References section. The protocol and informed consent we used for our interviews is provided in Appendix C, and the list of interview participants is provided in Appendix D. Our interview methods were reviewed by RAND's Human Subjects Protection Committee, and this project was assessed as not involving human subjects.[4] RAND's Human Subjects Protection Committee serves as RAND's IRB.

We conducted 15 interviews and consultations with experts. We reference the interviews but removed information that would identify which interview participant contributed the information (consistent with what we said we would do during the consent process). The identities of participants and their corresponding interview notes are accessible only to our team.

Limitations on Methodology

Our methodology included limitations that restrict our ability to draw certain conclusions. First, publication and citation are lagging indicators. The most-cited articles might not directly relate to importance, and recent publications may not yet have obtained as many citations. As a result, we may have missed critical topics that did not appear in our initial literature review. Accordingly, we augmented that review with additional searches of journal articles and gray literature on ethics and science. As noted above, our conversations with experts both led us to additional literature and aided our interpretation of what we read.

Second, we did not encounter the same quantity of international discussion of ethics in scientific research as that within the United States. Many international orga-

---

[4] RAND Human Subjects Protection Committee case number: 2017-0688-AM01.

nizations create their own reports or codes of ethics, which we reviewed and included in Appendix F or References, but this literature does not identify ethical differences among country members, or within those countries. The public discussion of different principles of ethics in specific countries is uneven and largely absent from the literature we reviewed. Therefore, our discussion on the country differences relating to research ethics is largely based on our interviews and is limited to the viewpoints of the interview participants that we could include in this study. A greater number of interviews could have resulted in a greater, more-detailed, and/or more-nuanced understanding of cultural and national differences.

Codebook

This codebook was used to identify relevant excerpts from each of the documents in our literature review. Each "code" is provided with its definition. These coding definitions do not necessarily match the ethical principle definitions (in Chapter Two) because the results of our analysis informed that chapter.

Training

The obligation falls on the researcher to be knowledgeable about ethical, legal, and regulatory requirements in their own country and international requirements for their discipline.

Monitoring

Includes IRBs and other types of monitoring bodies and the protocols they follow. Includes grievance mechanisms.

Compliance

This category includes names of specific laws, regulations, treaties, etc. Includes that publications (journals) have an obligation to not publish research that does not comply with codes of ethics. Includes legal compliance.

Remediation

Discussion of how ethical incidents should be responded to.

History

This includes history of how codes came to be developed, such as discussion of significant events that led to changes to ethics in research.

Adherence to Code

A statement committing the beneficiary of the code to its tenets.

Beneficence and Nonmaleficence

Beneficence is a concept in research ethics that in any research study, researchers should have the welfare of the research participant in mind as a goal. It often appears in tandem with nonmaleficence. *Maleficence* is considered the antonym of beneficence—it describes practices that decrease the welfare of the research participant. *Nonmaleficence* is not harming, or inflicting the least harm possible, to reach a beneficial outcome. Includes discussion of beneficence and nonmaleficence for humans, for animals, and for the environment or ecosystem.

Bribery

The act or practice of "money or favor given or promised in order to influence the judgment or conduct of a person in a position of trust."[1]

Conflict of Interest

"A conflict between the private interests and the official responsibilities of a person in a position of trust."[2]

Data Management

Includes discussion about sharing data (and data transparency) so other researchers can assess or reproduce the research; data handling; and how researchers choose which software tools to use.

---

[1]   *Bribe*, dictionary entry, Merriam-Webster online, undated,

[2]   *Conflict of interest*, dictionary entry, Merriam-Webster online, undated.

Duty to Society

A general principle that all those covered by the code "have the responsibility to contribute from their sphere of professional competence to the general well-being of society."[3]

Informed Consent

Informed consent is a voluntary agreement to participate in research. It is the process in which the subject has an understanding of the research and its risks and voluntarily agrees to participate.

Integrity

Discussion of upholding moral values. Includes professional conduct and plagiarism.

Report Results Accurately

Researchers are obligated to report results and data accurately.

Nondiscrimination

This principle ensures a zero-tolerance policy for discrimination based on race, gender, religion, and other demographics or group characteristics.

Nonexploitation

This principle prohibits personal gain or using research unfairly for one's own advantage.

Privacy and Confidentiality

Privacy refers to an individual's right to control access to their personal information, but it also includes access to their body (such as collection of biological specimens). Privacy is a subject's ability to control how other people see, touch, or obtain information about the subject. "'Confidentiality' refers to how private information provided by

---

[3]   American Society for Clinical Laboratory Science, *Code of Ethics*, undated.

individuals will be protected by the researcher from release. Describing just how the confidentiality of research information will be maintained is an important aspect of the consent process. Confidentiality is an extension of the concept of privacy; it refers to the subject's understanding of, and agreement to, the ways identifiable information will be stored and shared. Identifiable information can be printed information, electronic information, or visual information such as photographs."[4]

Professional Competence

Refers to researchers engaging only in work that they are qualified to perform while also participating in training and betterment programs with the intent of improving their skill sets. *Professional competence* differs from *professional discipline* in that the latter refers to researchers preserving the integrity and prestige of their fields, whereas the former focuses on the responsibility of the researcher to engage in research they are qualified for, as well as to pursue self-betterment. Includes discussion about choosing research methods, statistical methods, and sample sizes that are appropriate and would not cause misleading results.

Professional Discipline

Commitment to engaging in safe, sound research practices and "assist[ing] colleagues entering the profession by sharing knowledge and understanding of the ethics, responsibilities and needed competencies of their chosen area of research and practice."[5]

Other

This is a "miscellaneous" category for anything that does not fit elsewhere but that the coder wants to capture.

---

[4]   University of Pittsburgh Institutional Review Board, "IRB Guidance: Privacy Versus Confidentiality," April 1, 2014.

[5]   Association of Clinical Research Professionals, "Code of Ethics," undated.

Interview Informed Consent and Protocol

The following informed consent was sent to all interviewees by email prior to the interview. The interviews followed a semistructured format—meaning that the questions listed in this protocol were used as a starting-off point for discussions, but the interviewers had freedom to ask follow-up questions that may have led to differing lines of questioning.

Informed Consent

RAND Corporation is researching codes of conduct, ethical guidelines, and enforcement mechanisms that guide research and development (R&D) into various scientific and technical disciplines. We would like to interview you to understand specific ethical guidelines, how they work, and how they are implemented and enforced in your area of expertise.

This interview is voluntary. You may decline to participate in this interview, or you may decline to answer any questions during the interview. Our team will take notes, and we may use your responses to inform our research. In our report, we will include a list of everyone we have interviewed, and we may attribute your comments to you. If you do not consent to being identified in our report, please let us know.

This project is sponsored by the U.S. Government under the Intelligence Advanced Research Projects Activity (IARPA). If you have any questions or concerns about this project, please contact the RAND project investigators:

Cortney Weinbaum
[Contact information provided]

Eric Landree
[Contact information provided]

Interview Protocol

1. Is it okay for us to include your name and position in our report?
2. Please describe your current position and role.

3. Which best describes your relationship with codes of conduct or ethics: (1) you create, modify, or enforce codes of ethics, (2) you are a researcher who uses a code of ethic to guide your research, or (3) you research how ethics are used?

4. What are the primary [tenets] of ethics in your discipline?
    a. How do those [tenets] differ across international borders?

5. How are codes of conduct or ethics developed for your discipline?
    a. How do those codes cross international borders?
    b. Can you point us to a copy somewhere?

6. How are revisions made when new ethical issues arise?
    a. What events led to changes in the code of conduct (e.g. a crisis occurred; people died) or a response to the natural evolution of the discipline?
    b. Who decides what codes of ethics are used? What decision making criteria [are] used to decide whether to adopt a change (e.g. consensus, democratic vote, edict, etc.)?

7. What requirements or incentives do researchers have to follow the code of conduct?
    a. What risks do they incur by failing to follow the code of conduct?
    b. Who or what entity imposes these requirements, incentives, or disincentives?

8. How is the code of conduct enforced?
    a. Within one country versus across international borders?
    b. Who are the "enforcers"?
    c. What is your role in enforcement?

9. Is there anything else we should be aware of?

List of Interview Participants

We thank everyone who participated in our interviews and who shared their subject-matter expertise, including those who chose not to be identified here. We valued their candor and appreciated the insights they shared. Their professional experiences filled in vital details where our literature review was incomplete. The persons who remain unnamed are experts in the field of international scientific ethics, and they asked that their names and, in two cases, their organizations be withheld to protect the international scientific relationships they discussed with us.

This list is not in the same order as the interview numbers in footnotes throughout this report. We ask readers to not make assumptions about which interviewee may be associated with each interview number, as such assumptions may result in misleading conclusions. In alphabetical order:

- Anonymous government official in HHS
- Anonymous government officials at the U.S. Department of State
- Jeremy Berg, Ph.D., Editor-in-Chief of the *Science* family of journals
- Stephanie J. Bird, Ph.D., co-Editor-in-Chief of the *Journal of Science and Engineering Ethics*
- Diana Bowman, Ph.D., Senior Sustainability Scholar, Julie Ann Wrigley Global Institute of Sustainability
- Raja Chatila, Ph.D., Chair, IEEE Global Initiative on Ethics of Autonomous and Intelligent Systems and Director of the Institute of Intelligent Systems and Robotics at Sorbonne Université
- 从亚丽Yali Cong, Ph.D., Professor of Medical Ethics, Deputy Director of the Public Education Department of the Peking University School of Medicine, Associate Dean of the Institute of Medical Humanities
- C. K. Gunsalus, J.D., Director of the National Center for Professional and Research Ethics, Professor Emerita of Business, and Research Professor at the Coordinated Sciences Laboratory at the University of Illinois, Urbana Champaign
- Andrew M. Hebbeler, Ph.D., Deputy Director, Office of Science and Technology Cooperation, U.S. Department of State

- Rachelle Hollander, Ph.D., retired Program Director of the National Science Foundation Societal Dimensions of Engineering, Science, and Technology program and retired Program Director of the Center for Engineering, Ethics, and Society at the National Academy of Engineering
- Elsa Kania, Adjunct Fellow at the Center for a New American Security and co-founder of China Cyber and Intelligence Studies Institute
- Isidoros Karatzas, Head of Research Ethics and Integrity at the European Commission
- Annie Kersting, Ph.D., Director for University Relations and Science Education and Research Integrity Officer at Lawrence Livermore National Laboratory
- Gary Marchant, Ph.D., J.D., Regents' Professor and Lincoln Professor of Emerging Technologies, Law and Ethics, Sandra Day O'Connor College of Law at Arizona State University
- Anne Petersen, Ph.D., Professor at the University of Michigan and Chair of the Policy and Global Affairs Committee at the National Academies of Science, Engineering, and Medicine
- Edward You, Supervisory Special Agent in the FBI Weapons of Mass Destruction Directorate.

Bibliography of Literature

Altman, D. G., "Statistics and Ethics in Medical Research, III: How Large a Sample? *British Medical Journal*, Vol. 281, No. 6251, 1980, pp. 1336–1338.

Altman, D. G., "The Scandal of Poor Medical Research: We Need Less Research, Better Research, and Research Done for the Right Reasons," *British Medical Journal*, Vol. 308, No. 6924, 1994, pp. 283–284.

Altman, D. G., K. F. Schulz, D. Moher, M. Egger, F. Davidoff, D. Elbourne, P. C. Gøtzsche, and T. Lang, "The Revised CONSORT Statement for Reporting Randomized Trials: Explanation and Elaboration," *Annals of Internal Medicine*, Vol. 134, No. 8, 2001, pp. 663–694.

Angell, M., "The Ethics of Clinical Research in the Third World," *New England Journal of Medicine*, Vol. 337, No. 12, 1997, pp. 847–849.

Barbour, R. S., "Making Sense of Focus Groups," *Medical Education*, Vol. 39, No. 7, 2005, pp. 742–750.

Beecher, H. K., "Ethics and Clinical Research," *New England Journal of Medicine*, Vol. 274, No. 24, 1966, pp. 1354–1360.

Begley, C. G., and L. M. Ellis, "Drug Development: Raise Standards for Preclinical Cancer Research," *Nature*, Vol. 483, No. 7391, 2012, pp. 531–533.

Behrend, T. S., D. J. Sharek, A. W. Meade, and E. N. Wiebe, "The Viability of Crowdsourcing for Survey Research," *Behavior Research Methods*, Vol. 43, No. 3, 2011, pp. 800–813.

Bekelman, J. E., Y. Li, and C. P. Gross, "Scope and Impact of Financial Conflicts of Interest in Biomedical Research: A Systematic Review," *Journal of the American Medical Association*, Vol. 289, No. 4, 2003, pp. 454–465.

Brennan, T. A., D. J. Rothman, L. Blank, D. Blumenthal, S. C. Chimonas, J. J. Cohen, J. Goldman, J. P. Kassirer, H. Kimball, J. Naughton, and N. Smelser, "Health Industry Practices That Create Conflicts of Interest: A Policy Proposal for Academic Medical Centers," *Journal of the American Medical Association*, Vol. 295, No. 4, 2006, pp. 429–433.

Brey, P., and P. Jansen, "Ethics Assessment in Different Fields: Engineering Sciences," 2015.

Brown, C. A., and R. J. Lilford, "The Stepped Wedge Trial Design: A Systematic Review," *BMC Medical Research Methodology*, Vol. 6, 2006.

Button, K. S., J. P. A. Ioannidis, C. Mokrysz, B. A. Nosek, J. Flint, E. S. J. Robinson, and M. R. Munafò, "Power Failure: Why Small Sample Size Undermines the Reliability of Neuroscience," *Nature Reviews Neuroscience*, Vol. 14, No. 5, 2013, pp. 365–376.

Chalmers, I., "Underreporting Research Is Scientific Misconduct," *Journal of the American Medical Association*, Vol. 263, No. 10, 1990, pp. 1405–1408.

Chan, A. W., A. Hróbjartsson, M. T. Haahr, P. C. Gøtzsche, and D. G. Altman, "Empirical Evidence for Selective Reporting of Outcomes in Randomized Trials: Comparison of Protocols to Published Articles," *Journal of the American Medical Association*, Vol. 291, No. 20, 2004, pp. 2457–2465.

Chan, A. W., J. M. Tetzlaff, P. C. Gøtzsche, D. G. Altman, H. Mann, J. A. Berlin, K. Dickersin, A. Hróbjartsson, K. F. Schulz, W. R. Parulekar, K. Krleza-Jeric, A. Laupacis, and D. Moher, "SPIRIT 2013 Explanation and Elaboration: Guidance for Protocols of Clinical Trials," *British Medical Journal* (Clinical Research Edition), Vol. 346, No. 2, 2013.

Childress, J. F., R. R. Faden., R. D. Gaare, L. O. Gostin, J. Kahn, R. J. Bonnie, N. E. Kass, A. C. Mastroianni, J. D. Moreno, and P. Nieburg, "Public Health Ethics: Mapping the Terrain," *Journal of Law, Medicine and Ethics*, Vol. 30, No. 2, 2002, pp. 170–178.

Cios, K. J., and G. W. Moore, "Uniqueness of Medical Data Mining," *Artificial Intelligence in Medicine*, Vol. 26, No. 1–2, 2002, pp. 1–24.

Collins, F. S., E. D. Green, A. E. Guttmacher, and M. S. Guyer, "A Vision for the Future of Genomics Research," *Nature*, Vol. 422, No. 6934, 2003, pp. 835–847.

Colvin, J. G., "A Code of Ethics for Research in the 3rd World," *Conservation Biology*, Vol. 6, No. 3, 1992, pp. 309–311.

———, "Bridging the Gap: A Code of Ethics for Research in the Developing World," *Bioscience*, Vol. 43, No. 9, 1993, pp. 594–595.

Corbie-Smith, G., S. B. Thomas, M. V. Williams, and S. Moody-Ayers, "Attitudes and Beliefs of African Americans Toward Participation in Medical Research," *Journal of General Internal Medicine*, Vol. 14, No. 9, 1999, pp. 537–546.

Council for Big Data, Ethics, and Society, homepage, undated. As of March 14, 2019: http://bdes.datasociety.net/

Curtis, M. J., R. A. Bond, D. Spina, A. Ahluwalia, S. P. A. Alexander, M. A. Giembycz, A. Gilchrist, D. Hoyer, P. A. Insel, A. A. Izzo, A. J. Lawrence, D. J. Macewan, L. D. F. Moon, S. Wonnacott, A. H. Weston, and J. C. McGrath, "Experimental Design and Analysis and Their Reporting: New Guidance for Publication in *BJP*," *British Journal of Pharmacology*, Vol. 172, No. 14, 2015, pp. 3461–3471.

D'Agostino, R. B. Sr., J. M. Massaro., and L. M. Sullivan, "Non-Inferiority Trials: Design Concepts and Issues—The Encounters of Academic Consultants in Statistics," *Statistics in Medicine*, Vol. 22, No. 2, 2003, pp. 169–186.

DiCicco-Bloom, B., and B. F. Crabtree, "The Qualitative Research Interview," *Medical Education*, Vol. 40, No. 4, 2006, pp. 314–321.

Drummond, G. B., "Reporting Ethical Matters in the *Journal of Physiology*: Standards and Advice," *Journal of Physiology*, Vol. 587, No. 4, 2009, pp. 713–719.

Drumwright, M. E., "Company Advertising with a Social Dimension: The Role of Noneconomic Criteria," *Journal of Marketing*, Vol. 60, No. 4, 1996, pp. 71–87.

Easterbrook, P. J., R. Gopalan, J. A. Berlin, and D. R. Matthews, "Publication Bias in Clinical Research," *The Lancet*, Vol. 337, No. 8746, 1991, pp. 867–872.

Ellis, C., "Telling Secrets, Revealing Lives: Relational Ethics in Research with Intimate Others," *Qualitative Inquiry*, Vol. 13, No. 1, 2007, pp. 3–29.

Emanuel, E. J., D. Wendler, and C. Grady, "What Makes Clinical Research Ethical?" *Journal of the American Medical Association*, Vol. 283, No. 20, 2000, pp. 2701–2711.

Emanuel, E. J., D. Wendler, J. Killen, and C. Grady, "What Makes Clinical Research in Developing Countries Ethical? The Benchmarks of Ethical Research," *Journal of Infectious Diseases*, Vol. 189, No. 5, 2004, pp. 930–937.

Eysenbach, G., and J. E. Till, "Ethical Issues in Qualitative Research on Internet Communities," *British Medical Journal*, Vol. 323, No. 7321, 2001, pp. 1103–1105.

Fanelli, D., "How Many Scientists Fabricate and Falsify Research? A Systematic Review and Meta-Analysis of Survey Data," *PLoS ONE*, Vol. 4, No. 5, 2009.

Fang, F. C., R. G. Steen, and A. Casadevall, "Misconduct Accounts for the Majority of Retracted Scientific Publications," *Proceedings of the National Academy of Sciences of the United States of America*, Vol. 109, No. 42, 2012, pp. 17028–17033.

Ford, R. C., and W. D. Richardson, "Ethical Decision-Making—A Review of the Empirical Literature, *Journal of Business Ethics*, Vol. 13, No. 3, 1994, pp. 205–221.

Fossey, E., C. Harvey, F. McDermott, and L. Davidson, "Understanding and Evaluating Qualitative Research," *Australian and New Zealand Journal of Psychiatry*, Vol. 36, No. 6, 2002, pp. 717–732.

Freedman, B., "Equipoise and the Ethics of Clinical Research," *New England Journal of Medicine*, Vol. 317, No. 3, 1987, pp. 141–145.

Gallagher, T. H., A. D. Waterman, A. G. Ebers, V. J. Fraser, and W. Levinson, "Patients' and Physicians' Attitudes Regarding the Disclosure of Medical Errors," *Journal of the American Medical Association*, Vol. 289, No. 8, 2003, pp. 1001–1007.

Glickman, S. W., J. G. McHutchinson, E. D. Peterson, C. B. Cairns, R. A. Harrington, R. M. Califf, and K. A. Schulman, "Ethical and Scientific Implications of the Globalization of Clinical Research," *New England Journal of Medicine*, Vol. 360, No. 8, 2009, pp. 816–823.

Grunwald, A., "The Application of Ethics to Engineering and the Engineer's Moral Responsibility: Perspectives for a Research Agenda," *Science and Engineering Ethics*, Vol. 7, No. 3, 2001, pp. 415–428.

Guillemin, M., and L. Gillam, "Ethics, Reflexivity, and 'Ethically Important Moments' in Research," *Qualitative Inquiry*, Vol. 10, No. 2, 2004, pp. 261–280.

Hall, M. A., E. Dugan, B. Zheng, and A. K. Mishra, "Trust in Physicians and Medical Institutions: What Is It, Can It Be Measured, and Does It Matter?" *Milbank Quarterly*, Vol. 79, No. 4, 2001, pp. 613–639.

Harriss, D. J., and G. Atkinson, "International Journal of Sports Medicine: Ethical Standards in Sport and Exercise Science Research," *International Journal of Sports Medicine*, Vol. 30, No. 10, 2009, pp. 701–702.

———, "Update Ethical Standards in Sport and Exercise Science Research, *International Journal of Sports Medicine*, Vol. 32, No. 11, 2011, pp. 819–821.

———, "Ethical Standards in Sport and Exercise Science Research: 2014 Update," *International Journal of Sports Medicine*, Vol. 34, No. 12, 2013, pp. 1025–1028.

Hasson, F., S. Keeney, and H. McKenna, "Research Guidelines for the Delphi Survey Technique," *Journal of Advanced Nursing*, Vol. 32, No. 4, 2000, pp. 1008–1015.

Herkert, J. R., "Future Directions in Engineering Ethics Research: Microethics, Macroethics and the Role of Professional Societies," *Science and Engineering Ethics*, Vol. 7, No. 3, 2001, pp. 403–414.

———, "Engineering Research and Animal Subjects," lesson plan, 2004.

Higginson, I. J., and A. J. Carr, "Measuring Quality of Life: Using Quality of Life Measures in the Clinical Setting," *British Medical Journal*, Vol. 322, No. 7297, 2001, pp. 1297–1300.

Hooper, D. U., F. S. Chapin III, J. J. Ewel, A. Hector, P. Inchausti, S. Lavorel, J. H. Lawton, D. M. Lodge, M. Loreau, S. Naeem, B. Schmid, H. Setälä, A. J. Symstad, J. Vandermeer, and D. A. Wardle, "Effects of Biodiversity on Ecosystem Functioning: A Consensus of Current Knowledge," *Ecological Monographs*, Vol. 75, No. 1, 2005, pp. 3–35.

Huntingford, F. A., C. Adams, V. A. Braithwaite, S. Kadri, T. G. Pottinger, P. Sandøe, and J. F. Turnbull, "Current Issues in Fish Welfare," *Journal of Fish Biology*, Vol. 68, No. 2, 2006, pp. 332–372.

Jensen, P. B., L. J. Jensen, and S. Brunak, "Mining Electronic Health Records: Towards Better Research Applications and Clinical Care," *Nature Reviews Genetics*, Vol. 13, No. 6, 2012, pp. 395–405.

Kelley, K., B. Clark, V. Brown, and J. Sitzia, "Good Practice in the Conduct and Reporting of Survey Research," *International Journal for Quality in Health Care*, Vol. 15, No. 3, 2003, pp. 261–266.

Kuschel, R., "The Necessity of Ethical Codes in Research," *Psychiatry Today, Journal of the Yugoslav Psychiatric Association*, Vol. 2–3, 1998, pp. 147–174.

Leape, L. L., and D. M. Berwick, "Five Years After to Err Is Human: What Have We Learned?" *Journal of the American Medical Association*, Vol. 293, No. 19, 2005, pp. 2384–2390.

Lubchenco, J., "Entering the Century of the Environment: A New Social Contract for Science," *Science*, Vol. 279, No. 5350, 1998, pp. 491–497.

Ludvigsson, J. F., P. Otterblad-Olausson, B. U. Pettersson, and A. Ekbom, "The Swedish Personal Identity Number: Possibilities and Pitfalls in Healthcare and Medical Research," *European Journal of Epidemiology*, Vol. 24, No. 11, 2009, pp. 659–667.

Lurie, P., and S. M. Wolfe, "Unethical Trials of Interventions to Reduce Perinatal Transmission of the Human Immunodeficiency Virus in Developing Countries," *New England Journal of Medicine*, Vol. 337, No. 12, 1997, pp. 853–856.

Macaulay, A. C., L. E. Commanda, W. L. Freeman, N. Gibson, M. L. McCabe, C. M. Robbins, and P. L. Twohig, "Participatory Research Maximises Community and Lay Involvement," *British Medical Journal*, Vol. 319, No. 7212, 1999, pp. 774–778.

Macaulay, A. C., T. Delormier, A. M. McComber, E. J. Cross, L. P. Potvin, G. Paradis, R. L. Kirby, C. Saad-Haddad, and S. Desrosiers, "Participatory Research with Native Community of Kahnawake Creates Innovative Code of Research Ethics," *Canadian Journal of Public Health—Revue Canadienne De Santé Publique*, Vol. 89, No. 2, 1998, pp. 105–108.

Macklin, R., "Dignity Is a Useless Concept," *British Medical Journal*, Vol. 327, No. 7429, 2003, pp. 1419–1420.

Martinson, B. C., M. S. Anderson, and R. De Vries, "Scientists Behaving Badly," *Nature*, Vol. 435, No. 7043, 2005, pp. 737–738.

Matsuura, J. H., "Engineering Codes of Ethics: Legal Protection for Engineers," *The Bridge*, Vol. 47, No. 1, 2017, pp. 27–29.

Mayer, D. M., M. Kuenzi, R. Greenbaum, M. Bardes, and R. Salvador, "How Low Does Ethical Leadership Flow? Test of a Trickle-Down Model," *Organizational Behavior and Human Decision Processes*, Vol. 108, No. 1, 2009, pp. 1–13.

McCabe, D. L., L. K. Trevino, and K. D. Butterfield, "Cheating in Academic Institutions: A Decade of Research," *Ethics and Behavior*, Vol. 11, No. 3, 2001, pp. 219–232.

McGrath, J. C., G. B. Drummond, E. M. McLachlan, C. Kilkenny, and C. L. Wainwright, "Guidelines for Reporting Experiments Involving Animals: The ARRIVE Guidelines," *British Journal of Pharmacology*, Vol. 160, No. 7, 2010, pp. 1573–1576.

Minkler, M., "Community-Based Research Partnerships: Challenges and Opportunities," *Journal of Urban Health*, Vol. 82 (SUPPL. 2), 2005, pp. ii3–ii12.

Monzon, J. E., and A. Monzon-Wyngaard, "Ethics and Biomedical Engineering Education: The Continual Defiance," *2009 Annual International Conference of the IEEE Engineering in Medicine and Biology Society*, Vols. 1–20, 2009, pp. 2011–2014.

Morrow, V., and M. Richards, "The Ethics of Social Research with Children: An Overview," *Children and Society*, Vol. 10, No. 2, 1996, pp. 90–105.

Mouton, F., M. M. Malan, K. K. Kimppa, and H. S. Venter, "Necessity for Ethics in Social Engineering Research," *Computers and Security*, Vol. 55, 2015, pp. 114–127.

Newberry, B., "The Dilemma of Ethics in Engineering Education," *Science and Engineering Ethics*, Vol. 10, No. 2, 2004, pp. 343–351.

O'Fallon, M. J., and K. D. Butterfield, "A Review of the Empirical Ethical Decision-Making Literature: 1996–2003," *Journal of Business Ethics*, Vol. 59, No. 4, 2005, pp. 375–413.

Paasche-Orlow, M. K., H. A. Taylor, and F. L. Brancati, "Readability Standards for Informed-Consent Forms as Compared with Actual Readability," *New England Journal of Medicine*, Vol. 348, No. 8, 2003, pp. 721–726.

Peterson, J., S. Garges, M. Giovanni, P. McInnes, L. Wang, J. A. Schloss, V. Bonazzi, J. E. McEwen, K. A. Wetterstrand, C. Deal, C. C. Baker, V. Di Francesco, T. K. Howcroft, R. W. Karp, R. D. Lunsford, C. R. Wellington, T. Belachew, M. Wright, C. Giblin, H. David, M. Mills, R. Salomon, C. Mullins, B. Akolkar, L. Begg, C. Davis, L. Grandison, M. Humble, J. Khalsa, A. Roger Little, H. Peavy, C. Pontzer, M. Portnoy, M. H. Sayre, P. Starke-Reed, S. Zakhari, J. Read, B. Watson, and M. Guyer, "The NIH Human Microbiome Project," *Genome Research*, Vol. 19, No. 12, 2009, pp. 2317–2323.

Portaluppi, F., M. H. Smolensky, and Y. Touitou, "Ethics and Methods for Biological Rhythm Research on Animals and Human Beings," *Chronobiology International*, Vol. 27, 9–10, 2010, pp. 1911–1929.

Rees, D. A., and J. C. Alcolado, "Animal Models of Diabetes Mellitus," *Diabetic Medicine*, Vol. 22, No. 4, 2005, pp. 359–370.

Rennie, D., V. Yank, and L. Emanuel, "When Authorship Fails: A Proposal to Make Contributors Accountable," *Journal of the American Medical Association*, Vol. 278, No. 7, 1997, pp. 579–585.

Roden, D. M., J. M. Pulley, M. A. Basford, G. R. Bernard, E. W. Clayton, J. R. Balser, and D. R. Masys, "Development of a Large-Scale De-identified DNA Biobank to Enable Personalized Medicine," *Clinical Pharmacology and Therapeutics*, Vol. 84, No. 3, 2008, pp. 362–369.

Rossi, S., M. Hallet, P. M. Rossini, A. Pascual-Leone, G. Avanzini, S. Bestmann, A. Berardelli, C. Brewer, T. Canli, R. Cantello, R. Chen, J. Classen, M. Demitrack, V. Di Lazzaro, C. M. Epstein, M. S. George, F. Fregni, R. Ilmoniemi, R. Jalinous, B. Karp, J. P. Lefaucheur, S. Lisanby, S. Meunier, C. Miniussi, P. Miranda, F. Padberg, W. Paulus, A. Peterchev, C. Porteri, M. Provost, A. Quartarone, A. Rotenberg, J. Rothwell, J. Ruohonen, H. Siebner, G. Thut, J. Valls-Solè, V. Walsh, Y. Ugawa, A. Zangen, and U. Ziemann, "Safety, Ethical Considerations, and Application Guidelines for the Use of Transcranial Magnetic Stimulation in Clinical Practice and Research," *Clinical Neurophysiology*, Vol. 120, No. 12, 2009, pp. 2008-2039.

Rothman, K. J., and K. B. Michels, "The Continuing Unethical Use of Placebo Controls," *New England Journal of Medicine*, Vol. 331, No. 6, 1994, pp. 394–398.

Rothwell, P. M., "External Validity of Randomised Controlled Trials: "To Whom Do the Results of This Trial Apply?" *The Lancet*, Vol. 365, No. 9453, 2005, pp. 82–93.

Runeson, P., and M. Host, "Guidelines for Conducting and Reporting Case Study Research in Software Engineering," *Empirical Software Engineering*, Vol. 14, No. 2, 2009, pp. 131–164.

Shendure, J., R. D. Mitra, C. Varma, G. M. Church, "Advanced Sequencing Technologies: Methods and Goals," *Nature Reviews Genetics*, Vol. 5, No. 5, 2004, pp. 335–344.

Shewan, L. G., and A. J. S. Coats, "Ethics in the Authorship and Publishing of Scientific Articles," *International Journal of Cardiology*, Vol. 144, No. 1, 2010, pp. 1–2.

Shuman, L. J., S. M. Besterfield-Sacre, and J. McGourty, "The ABET 'Professional Skills': Can They Be Taught? Can They Be Assessed?" *Journal of Engineering Education*, Vol. 94, No. 1, 2005, pp. 41–55.

Sikes, R. S., and W. L. Gannon, "Guidelines of the American Society of Mammalogists for the Use of Wild Mammals in Research," *Journal of Mammalogy*, Vol. 92, No. 1, 2011, pp. 235–253.

Singer, J., and N. G. Vinson, "Ethical Issues in Empirical Studies of Software Engineering," *IEEE Transactions on Software Engineering*, Vol. 28, No. 12, 2002, pp. 1171–1180.

Smith, H. J., T. Dinev, and H. Xu, "Information Privacy Research: An Interdisciplinary Review," *MIS Quarterly: Management Information Systems*, Vol. 35, No. 4, 2011, pp. 989–1015.

Stelfox, H. T., G. Chua., K. O'Rourke, and A. S. Detsky, "Conflict of Interest in the Debate over Calcium-Channel Antagonists," *New England Journal of Medicine*, Vol. 338, No. 2, 1998, pp. 101–106.

Stewart, K. A., and A. H. Segars, "An Empirical Examination of the Concern for Information Privacy Instrument," *Information Systems Research*, Vol. 13, No. 1, 2002, pp. 36–49.

Sung, N. S., W. F. Crowley, Jr., M. Genel, P. Salber, L. Sandy, L. M. Sherwood, S. B. Johnson, V. Catanese, H. Tilson, K. Getz, E. L. Larson, D. Scheinberg, E. A. Reece, H. Slavkin, A. Dobs, J. Grebb, R. A. Martinez, A. Korn, and D. Rimoin, "Central Challenges Facing the National Clinical Research Enterprise," *Journal of the American Medical Association*, Vol. 289, No. 10, 2003, pp. 1278–1287.

Tarullo, A. R., and M. R. Gunnar, "Child Maltreatment and the Developing HPA Axis," *Hormones and Behavior*, Vol. 50, No. 4, 2006, pp. 632–639.

Tenopir, C., S. Allard, K. Douglass, A. U. Aydinoglu, L. Wu, E. Read, M. Manoff, and M. Frame, "Data Sharing by Scientists: Practices and Perceptions," *PLoS ONE*, Vol. 6, No. 6, 2011.

Thabane, L., J. Ma, R. Chu, J. Cheng, A. Ismaila, L. P. Rios, R. Robson, M. Thabane, L. Giangregorio, and C. H. Goldsmith, "A Tutorial on Pilot Studies: The What, Why and How," *BMC Medical Research Methodology*, Vol. 10, 2010.

Thomas, S. B., and S. C. Quinn, "Public Health Then and Now: The Tuskegee Syphilis Study, 1932 to 1972: Implications for HIV Education and AIDS Risk Education Programs in the Black Community," *American Journal of Public Health*, Vol. 81, No. 11, 1991, pp. 1498–1504.

Thompson, D. F., "Understanding Financial Conflicts of Interest," *New England Journal of Medicine*, Vol. 329, No. 8, 1993, pp. 573–576.

Thuret, S., L. D. Moon, and F. H. Gage, "Therapeutic Interventions After Spinal Cord Injury," *Nature Reviews Neuroscience*, Vol. 7, No. 8, 2006, pp. 628–643.

Tracy, S. J., "Qualitative Quality: Eight "Big-Tent" Criteria for Excellent Qualitative Research," *Qualitative Inquiry*, Vol. 16, No. 10, 2010, pp. 837–851.

Treviño, L. K., K. D. Butterfield, and D. L. McCabe, "The Ethical Context in Organizations: Influences on Employee Attitudes and Behaviors," *Business Ethics Quarterly*, Vol. 8, No. 3, 1998, pp. 447–476.

Tunis, S. R., D. B. Stryer, and C. M. Clancy, "Practical Clinical Trials: Increasing the Value of Clinical Research for Decision Making in Clinical and Health Policy," *Journal of the American Medical Association*, Vol. 290, No. 12, 2003, pp. 1624–1632.

Walsham, G., "Doing Interpretive Research," *European Journal of Information Systems*, Vol. 15, No. 3, 2006, pp. 320–330.

Williams, J. R., "The Declaration of Helsinki and Public Health," *Bulletin of the World Health Organization*, Vol. 86, No. 8, 2008, pp. 650–652.

Wolf, S. M., F. P. Lawrenz, C. A. Nelson, J. P. Kahn, M. K. Cho, E. W. Clayton, J. G. Fletcher, M. K. Georgieff, D. Hammerschmidt, K. Hudson, J. Illes, V. Kapur, M. A. Keane, B. A. Koenig, B. S. LeRoy, E. G. McFarland, J. Paradise, L. S. Parker, S. F. Terry, B. Van Ness, B. S. Wilfond, "Managing Incidental Findings in Human Subjects Research: Analysis and Recommendations," *Journal of Law, Medicine and Ethics*, Vol. 36, No. 2, 2008, pp. 219–248.

Wynne, B., "Creating Public Alienation: Expert Cultures of Risk and Ethics on GMOs," *Science as Culture*, Vol. 10, No. 4, 2001, pp. 445–481.

Zimmermann, M., "Ethical Guidelines for Investigations of Experimental Pain in Conscious Animals," *Pain*, Vol. 16, No. 2, 1983, pp. 109–110.

Bibliography of Codes of Conduct

Academy of Management, "AOM Code of Ethics," December 6, 2017. As of March 18, 2019: http://aom.org/About-AOM/AOM-Code-of-Ethics.aspx?terms=code%20of%20conduct

ACM—*See* Association for Computing Machinery.

American Anthropological Association, "Principles of Professional Responsibility," November 1, 2012. As of February 28, 2019: http://ethics.americananthro.org/ethics-statement-0-preamble/

American Association of Physicists in Medicine, *Code of Ethics for the American Association of Physicists in Medicine: Report of Task Group 109*, 2009.

American Astronomical Society, "AAS Ethics Statement," 2010.

American Geophysical Union, "AGU Scientific Integrity and Professional Ethics," 2017.

American Geosciences Institute, "American Geosciences Institute Guidelines for Ethical Professional Conduct," 2015.

American Mathematical Society, "Policy Statement on Ethical Guidelines," 2005.

American Medical Association, "Code of Medical Ethics of the American Medical Association: Opinions on Research and Innovation," 2016.

American Physical Society, "APS Guidelines for Professional Conduct," 2002.

American Psychological Association, "American Psychological Association's Ethical Principles of Psychologists and Code of Conduct," 2017.

American Society for Biochemistry and Molecular Biology, "Code of Ethics," 2018.

American Society for Clinical Laboratory Science, "Code of Ethics," undated.

American Society for Microbiology, "Code of Ethics," 2005.

American Society of Human Genetics, "Code of Ethics," 2017.

American Society of Mechanical Engineers, "Code of Ethics," 1998.

American Sociological Association, "The Code of Ethics of the American Sociological Association," 2018.

American Statistical Association, "Ethical Guidelines for Statistical Practice," 2018.

APA—*See* American Psychological Association.

ASM—*See* American Society for Microbiology.

Association for Computing Machinery, "ACM Code of Ethics and Professional Conduct," 1992.

———, "ACM Code of Ethics and Professional Conduct," 2018.

Association of Clinical Research Professionals, "Code of Ethics," undated.

Biomedical Engineering Society, "Code of Ethics," 2004.

British Computing Society, "Code of Conduct for BCS Members," 2015.

Coats, A. J. S., "Ethical Authorship and Publishing," *International Journal of Cardiology*, Vol. 131, No. 2, 2009, pp. 149–150.

Ecological Society of America, "ESA Code of Ethics," 2013.

Institute for Operations Research and the Management Sciences, "INFORMS Ethics Guidelines," 2018.

International Society for Environmental Epidemiology, "Ethics Guidelines for Environmental Epidemiologists," 2012.

International Society of Ethnobiology, "ISE Code of Ethics," 2008.

International Sociological Association, "Code of Ethics," 2001.

National Academy of Sciences, National Academy of Engineering, and Institute of Medicine of the National Academies, *On Being a Scientist: A Guide to Responsible Conduct in Research*, 3rd Edition, 2009.

National Society of Professional Engineers, *Code of Ethics*, Alexandria, Va., 2018. As of February 26, 2019:
https://www.nspe.org/resources/ethics/code-ethics

Papademas, D., "IVSA Code of Research Ethics and Guidelines," *Visual Studies*, Vol. 24, No. 3, 2009, pp. 250–257.

Society for Applied Anthropology, "Statement of Ethics and Professional Responsibilities," 2015.

Society for Neuroscience, "Global Statement on the Use of Animals in Research," 2018a.

———, "Policies on the Use of Animals and Humans in Research," 2018b.

———, "Research Practices for Scientific Rigor: A Resource for Discussion, Training, and Practice," 2018c.

Society of Toxicology, "Code of Ethics," 2012.

WMA—*See* World Medical Association.

World Medical Association, "DECLARATION OF HELSINKI: Recommendations Guiding Doctors in Clinical Research," 18th World Medical Assembly, Helsinki, Finland, June 1964.

———, "Declaration of Helsinki: Recommendations Guiding Physicians in Biomedical Research Involving Human Subjects," *Journal of the American Medical Association*, Vol. 277, No. 11, March 19, 1997. As of February 26, 2019:
https://jamanetwork.com/journals/jama/articlepdf/414713/jama_277_11_038.pdf

———, "Declaration of Helsinki: Ethical Principles for Medical Research Involving Human Subjects," 2000. As of March 18, 2019:
https://www.scopus.com/inward/record.uri?eid=2-s2.0-0034694856&partnerID=40&md5=4cb3df2cd8f84e5f9d20bd797f6a4e7d; https://jamanetwork.com/journals/jama/articlepdf/193387/JSC00472.pdf

———, "Declaration of Helsinki: Ethical Principles for Medical Research Involving Human Subjects, *Journal of the American Medical Association*, Vol. 310, No. 20, 2013. As of February 26, 2019:
https://jamanetwork.com/journals/jama/articlepdf/1760318/jsc130006.pdf

References

Accreditation Board of Engineering and Technology, homepage, 2018. As of March 8, 2019:
https://www.abet.org/

"AI4EU Project," webpage, European Union, 2018. As of March 7, 2019:
http://ai4eu.org/

Alberts, B., M. W. Kirschner, S. Tilghman, and H. Varmus, "Rescuing US Biomedical Research from its Systemic Flaws," *Proceedings of the National Academy of Sciences of the United States of America*, Vol. 111, No. 16, 2014, pp. 5773–5777.

American Association for the Advancement of Science, "AAAS, China, and Ethics in Science." 2018. As of March 8, 2019:
https://www.aaas.org/programs/scientific-responsibility-human-rights-law/
aaas-china-and-ethics-science

American Chemical Society, "Materials for Ethics Education," webpage, 2018a. As of March 8, 2019:
https://www.acs.org/content/acs/en/about/governance/committees/ethics/ethics-case-studies.html

———, "Standards, Guidelines, and ACS Approval Process," webpage, 2018b. As of March 8, 2019:
https://www.acs.org/content/acs/en/education/policies.html

American Medical Association, "Code of Medical Ethics: Preface and Preamble," 2016. As of February 26, 2019:
https://www.ama-assn.org/delivering-care/code-medical-ethics-preface-preamble

———, "Principles of Medical Ethics," 2018. As of February 28, 2019:
https://www.ama-assn.org/delivering-care/ama-principles-medical-ethics

American Psychological Association, *Position on Ethics and Interrogation*, Washington, D.C., 2019a. As of February 27, 2019:
https://www.apa.org/ethics/programs/position/

———, *Timeline of APA Policies and Actions Related to Detainee Welfare and Professional Ethics in the Context of Interrogation and National Security*, Washington, D.C., 2019b. As of February 27, 2019:
https://www.apa.org/news/press/statements/interrogations.aspx

APA—*See* American Psychological Association.

Bajpai, V., "Rise of Clinical Trials Industry in India: An Analysis," *ISRN Public Health*, Vol. 2013, Article 167059, 2013.

Belluz, J., "20 Years Ago, Research Fraud Catalyzed the Anti-Vaccination Movement. Let's Not Repeat History," Vox.com, 2018. As of March 18, 2019:
https://www.vox.com/2018/2/27/17057990/andrew-wakefield-vaccines-autism-study

Berman, M., J. Jouvenal, and A. Selk, "Authorities Used DNA, Genealogy Website to Track Down 'Golden State Killer' Suspect Decades After Crimes," *Washington Post*, April 26, 2018.

Berti Suman, A., and R. Pierce, "Challenges for Citizen Science and the EU Open Science Agenda Under the GDPR," *European Data Protection Law Review*, Vol. 4, No. 3, 2018.

Canadian Panel on Research Ethics, *Tri-Council Policy Statement: Ethical Conduct for Research Involving Humans*, Ottawa, Canada: Government of Canada, 2014. As of March 7, 2019: http://www.pre.ethics.gc.ca/eng/policy-politique/initiatives/tcps2-eptc2/Default/

CBD—*See* Convention on Biological Diversity.

Centers for Disease Control and Prevention, "U.S. Public Health Service Syphilis Study at Tuskegee," 2015.

Chari, R., L. J. Matthews, M. S. Blumenthal, A. F. Edelman, and T. Jones, "The Promise of Community Citizen Science," Santa Monica, Calif.: RAND Corporation, PE-256-RC, 2017. As of March 18, 2019: https://www.rand.org/pubs/perspectives/PE256.html

Chennells, R., and Andries Steenkamp, "International Genomics Research Involving the San People," in Doris Schroeder, J. C. Cook, Francois Hirsch, Solveig Fenet, and Vasantha Muthuswamy, *Ethics Dumping, Case Studies from North-South Research Collaborations*, New York: Springer International, 2017.

CitiGen, "What Happens to Your Genetic Data When You Take a Commercial DNA Ancestry Test?" 2017. As of February 13, 2019: http://www.citigen.org/2017/07/12/what-happens-to-your-genetic-data-when-you-take-a-commercial-dna-ancestry-test/

Committee on Economic, Social, and Cultural Rights, "General Comment No. 20, Non-Discrimination in Economic, Social and Cultural Rights," 2009.

Consortium of Independent Review Boards, "CIRB Members," 2018. As of March 18, 2019: http://www.consortiumofirb.org/cirb-members/

Convention on Biological Diversity, "The Cartagena Protocol on Biosafety," 2003. As of March 7, 2019: http://bch.cbd.int/protocol

———, "History of the Convention," webpage, 2018a. As of March 7, 2019: https://www.cbd.int/history/

———, "The Convention on Biological Diversity," webpage, 2018b. As of March 7, 2019: https://www.cbd.int/convention/default.shtml

Cornell University, ArXiv, 2018. As of March 18, 2019: https://arxiv.org/

Davis, M., "Thinking Like an Engineer: The Place of a Code of Ethics in the Practice of a Profession, *Philosophy and Public Affairs*, Vol. 20, No. 2, 1991, pp. 150–167.

Denyer, S., "China Used to Harvest Organs from Prisoners. Under Pressure, the Practice Is Finally Ending," *Washington Post*, September 15, 2017. As of March 8, 2019: https://www.washingtonpost.com/world/asia_pacific/in-the-face-of-criticism-china-has-been-cleaning-up-its-organ-transplant-industry/2017/09/14/d689444e-e1a2-11e6-a419-eefe8eff0835_story.html?utm_term=.5a434f89ddd5

EC—*See* European Commission.

Electronic Privacy Information Center, "The Privacy Act of 1974," 2018. As of March 4, 2019:
https://epic.org/privacy/1974act/#history

European Commission, "Reducing the Risk of Exporting Non Ethical Practices to Third Countries," GARRI-6-2014, request for proposals, December 10, 2013. As of March 8, 2019:
http://ec.europa.eu/research/participants/portal/desktop/en/opportunities/h2020/topics/garri-6-2014.html

———, *Ethics for Researchers*, 2018a.

———, "A Global Code of Conduct to Counter Ethics Dumping," press release, June 17, 2018b. As of March 7, 2019:
http://ec.europa.eu/research/infocentre/article_en.cfm?id=/research/headlines/news/article_18_06_27_en.html?infocentre&item=Infocentre&artid=49377

———, "Open Science," webpage, 2018c. As of March 7, 2019:
https://ec.europa.eu/research/openscience/index.cfm

———, "Research and Innovation," webpage, 2018d. As of March 7, 2019:
https://ec.europa.eu/info/departments/research-and-innovation_en

EuroStemCell, "Regulation of Stem Cell Research in Europe," webpage, 2018. As of March 7, 2019:
https://www.eurostemcell.org/regulation-stem-cell-research-europe

European Union, *The Precautionary Principle*, Communication(2000) 1Final, February 2, 2000. As of March 7, 2019:
https://eur-lex.europa.eu/legal-content/EN/TXT/?uri=LEGISSUM%3Al32042

Executive Office of the President, President's Council of Advisors on Science and Technology, *Report to the President: Big Data and Privacy: A Technological Perspective*, May 2014. As of April 4, 2019:
https://obamawhitehouse.archives.gov/sites/default/files/microsites/ostp/PCAST/pcast_big_data_and_privacy_-_may_2014.pdf

Garnett, K., and D. J. Parsons, "Multi-Case Review of the Application of the Precautionary Principle in European Union Law and Case Law," *Risk Analysis*, Vol. 37, No. 3, 2017, pp. 502–516.

Google, "Google AI," tools, 2018. As of March 18, 2019:
https://ai.google/tools/

Gunsalus, C. K., and A. Robinson, "Nine Pitfalls of Research Misconduct," *Nature*, Vol. 557, 2018, pp. 297–299. As of May 16, 2018:
https://www.nature.com/articles/d41586-018-05145-6

Gunsalus, C. K., A. R. Marcus, and I. Oransky, "Institutional Research Misconduct Reports Need More Credibility," *JAMA—Journal of the American Medical Association*, Vol. 319, No. 13, 2018, pp. 1315–1316.

Gustafsson, Bengt, Lars Rydén, Gunnar Tybell, and Peter Wallensteen, "Focus on: The Uppsala Code of Ethics for Scientists," *Journal of Peace Research*, Vol. 21, No. 4, 1984, pp. 311–316. As of February 26, 2019:
https://www.jstor.org/stable/pdf/423746.pdf?refreqid=excelsior%3A3f6ab2d34d88ac1e5c5df517894240d3

Hackett, D. W., "16 Year Old 'Vaccines Cause Autism' Paper Withdrawn, Finally," 2018. As of March 18, 2019:
https://www.precisionvaccinations.com/mmr-vaccine-does-not-cause-autism-says-cdc

Halpin, R., "Can Unethically Produced Data Be Used Ethically?" *Medicine and Law*, Vol. 29, 2010, pp. 373–387.

HHS—*See* U.S. Department of Health and Human Services.

Holloway, D. T., "Regulating Amateurs," *The Scientist*, 2013.

IEEE—*See* Institute of Electrical and Electronics Engineers.

Institute of Electrical and Electronics Engineers, "The IEEE Global Initiative on Ethics of Autonomous and Intelligent Systems," 2017. As of April 4, 2019:
https://standards.ieee.org/industry-connections/ec/autonomous-systems.html

Interacademy Partnership, *Doing Global Science: A Guide to Responsible Conduct in the Global Research Enterprise*, Washington, D.C.: U.S. National Academies of Science, Engineering, and Medicine, 2016. As of March 8, 2019:
http://www.interacademies.org/33345/Doing-Global-Science-A-Guide-to-Responsible-Conduct-in-the-Global-Research-Enterprise

International Council for Harmonisation, "ICH Guidelines," undated. As of April 4, 2019:
https://www.ich.org/products/guidelines.html

Joffe, S., and H. F. Lynch, "Federal Right-to-Try Legislation: Threatening the FDA's Public Health Mission," *New England Journal of Medicine*, Vol. 378, No. 8, 2018, pp. 695–697.

Johnson, C. Y., "Project 'Gaydar,'" *Boston Globe*, 2009.

Khan, N. R., H. Saad, C. S. Oravec, N. Rossi, V. Nguyen, G. T. Venable, J. C. Lillard, P. Patel, D. R. Taylor, B. N. Vaughn, D. Kondziolka, F. G. Barker, L. M. Michael, and P. Klimo, "A Review of Industry Funding in Randomized Controlled Trials Published in the Neurosurgical Literature: The Elephant in the Room, *Neurosurgery*, Vol. 83, No. 5, 2018, pp. 890–897.

Kloumann, I., and J. Kleinberg, "Community Membership Identification from Small Seed Sets," Cornell University, 2014. As of March 18, 2019:
http://www.cs.cornell.edu/home/kleinber/kdd14-seed.pdf

Kriebel, D., J. Tickner, P. Epstein, J. Lemons, R. Levins, E. L. Loechler, and M. Stoto, "The Precautionary Principle in Environmental Science," *Environmental Health Perspectives,* Vol. 109, No. 9, 2001, pp. 871–876.

Krubiner, C. B., and R. R. Faden, "Pregnant Women Should Not Be Categorised as a 'Vulnerable Population' in Biomedical Research Studies: Ending a Vicious Cycle of 'Vulnerability,'" *Journal of Medical Ethics*, Vol. 43, 2017, pp. 664–665.

Lurie, Peter, M.D., MPH, Associate Commissioner for Public Health Strategy and Analysis, U.S. Food and Drug Administration, statement before the U.S. Senate Committee on Homeland Security and Government Affairs, September 22, 2016.

Malički, M., and A. Marušić, "Is There a Solution to Publication Bias? Researchers Call for Changes in Dissemination of Clinical Research Results," *Journal of Clinical Epidemiology*, Vol. 67, No. 10, 2014, pp. 1103–1110.

Matosin, N., E. Frank, M. Engel, J. S. Lum, and K. A. Newell, "Negativity Towards Negative Results: A Discussion of the Disconnect Between Scientific Worth and Scientific Culture," *Disease Models and Mechanisms*, Vol. 7, No. 2, 2014, pp. 171–173.

Mercatus Center, "Permissionless Innovation," webpage, 2018. As of March 19, 2019:
http://permissionlessinnovation.org/

Millum, J., and C. Grady, "The Ethics of Placebo-Controlled Trials: Methodological Justifications," *Contemporary Clinical Trials*, Vol. 36, No. 2, 2013, pp. 510–514.

Mlinarić, A., M. Horvat, and V. Šupak Smolčić, "Dealing with the Positive Publication Bias: Why You Should Really Publish Your Negative Results, *Biochemia Medica*, Vol. 27, No. 3, 2017.

National Academies of Sciences, Engineering, and Medicine, *Fostering Integrity in Research*, Washington, D.C.: The National Academies Press, 2017.

National Human Genome Research Institute, "The Genetic Information Nondiscrimination Act of 2008," April 17, 2017. As of February 5, 2019:
https://www.genome.gov/27568492/the-genetic-information-nondiscrimination-act-of-2008/

National Institutes of Health, "About," webpage, 2018a. As of March 19, 2019:
https://allofus.nih.gov/about

———, "All of Us Research Program Protocol," webpage, 2018b. As of March 19, 2019:
https://allofus.nih.gov/about/all-us-research-program-protocol

———, "Vulnerable and Other Populations Requiring Additional Protections," webpage, last updated on January 7, 2019. As of February 28, 2019:
https://grants.nih.gov/policy/humansubjects/policies-and-regulations/vulnerable-populations.htm

National Nanotechnologies Initiative, "Ethical, Legal, and Societal Issues," webpage, 2018. As of March 8, 2019:
https://www.nano.gov/you/ethical-legal-issues

National Society of Professional Engineers, *Code of Ethics*, Alexandria, Va., 2018. As of February 26, 2019:
https://www.nspe.org/resources/ethics/code-ethics

Nordling, Linda, "San People of Africa Draft Code of Ethics for Researchers," *Science*, March 17, 2017.

Office of Science and Technology Policy, "Memorandum for the Heads of Executive Departments and Agencies," 2013.

Office of the Director of National Intelligence, "Public Access to IARPA Research," webpage, 2018. As of March 19, 2019:
https://www.iarpa.gov/index.php/working-with-iarpa/public-access-to-iarpa-research

Parliament of the United Kingdom, "UK Can Lead the Way on Ethical AI, Says Lords Committee," April 16, 2017. As of March 7, 2019:
https://www.parliament.uk/business/committees/committees-a-z/lords-select/ai-committee/news-parliament-2017/ai-report-published/

Pew Research Center, "Social Media Fact Sheet," fact sheet, 2018. As of February 28, 2019:
http://www.pewinternet.org/fact-sheet/social-media/

Public Library of Science, "PLOS Global Participation Initiative," FAQ, 2018. As of March 19, 2019:
https://www.plos.org/faq#loc-plos-global-participation-initiative

Pugwash, "Pugwash Conferences on Science and World Affairs," homepage, 2018. As of February 26, 2019:
https://pugwash.org/

Qiu, J., "Injection of Hope Through China's Stem-Cell Therapies," *Lancet Neurology*, Vol. 7, No. 2, 2008, pp. 122–123.

Quin, A., "Fraud Scandals Sap China's Dream of Becoming a Science Superpower," *New York Times*, October 13, 2017. As of February 6, 2019:
https://www.nytimes.com/2017/10/13/world/asia/china-science-fraud-scandals.html

Rahwan, I., "Society-in-the-Loop: Programming the Algorithmic Social Contract," *Ethics and Information Technology*, Vol. 20, No. 1, 2018, pp. 5–14.

Salem, D. N., and M. M. Boumil, "Conflict of Interest in Open-Access Publishing," *New England Journal of Medicine*, Vol. 369, No. 5, 2013, p. 491.

Schroeder, Doris, J. C. Cook, Francois Hirsch, Solveig Fenet, and Vasantha Muthuswamy, *Ethics Dumping, Case Studies from North-South Research Collaborations*, New York: Springer International, 2017.

Schwartz, A. E., "Engineering Society Codes of Ethics: A Bird's-Eye View," *The Bridge*, Vol. 47, No. 1, 2017, pp. 21–26.

Science and Environmental Health Network, "Wingspread Conference on the Precautionary Principle," webpage, 1998. As of March 7, 2019:
http://sehn.org/wingspread-conference-on-the-precautionary-principle/

Shah, S. K., J. Kimmelman, A. D. Lyerly, H. F. Lynch, F. G. Miller, R. Palacios, C. A. Pardo, and C. Zorrilla, "Bystander Risk, Social Value, and Ethics of Human Research," *Science*, Vol. 360, No. 6385, 2018, pp. 158–159.

Shapiro, E. "'I Honestly Never Thought They Would Find Him': DNA Test, Genetic Genealogy Lead to Arrest in Woman's 2001 Killing," ABC News, November 6, 2018.

Skierka, A., and K. Michels, "Ethical Principles and Placebo-Controlled Trials: Interpretation and Implementation of the Declaration of Helsinki's Placebo Paragraph in Medical Research," *BMC Medical Ethics*, 2018.

Smith, E., Salil Gunashekar, Sarah Parks, Catherine A. Lichten, Louise Lepetit, Molly Morgan Jones, Catriona Manville, and Calum MacLure, "Monitoring Open Science Trends in Europe," Santa Monica, Calif.: RAND Corporation, TL-252-EC, 2017. As of March 19, 2019:
https://www.rand.org/pubs/tools/TL252.html

Smith, E., Sarah Parks, Salil Gunashekar, Catherine A. Lichten, Anna Knack, and Catriona Manville, "Open Science: The Citizen's Role and Contribution to Research," Santa Monica, Calif.: RAND Corporation, PE-246-CI, 2017. As of March 19, 2019:
https://www.rand.org/pubs/perspectives/PE246.html

Society of Toxicology, "Code of Ethics and Conflict of Interest," webpage, last revised 2012. As of April 3, 2019:
https://www.toxicology.org/about/vp/code-of-ethics.asp

South African San Institute and the TRUST Project, "San Code of Research Ethics," Kimberley, South Africa, 2017. As of March 4, 2019:
http://trust-project.eu/wp-content/uploads/2017/03/San-Code-of-RESEARCH-Ethics-Booklet-final.pdf

"TRUST Project," homepage, European Union, 2018. As of March 7, 2019:
http://trust-project.eu/

U.S. Department of Health and Human Services, "International Compilation of Human Research Standards," 2018. As of March 7, 2019:
https://www.hhs.gov/ohrp/sites/default/files/2018-International-Compilation-of-Human-Research-Standards.pdf

U.S. Department of State, "Science, Technology, and Innovation Partnerships," webpage, 2018. As of March 8, 2019:
https://www.state.gov/e/oes/stc/partnerships/

U.S. Food and Drug Administration, "Institutional Review Boards Frequently Asked Questions: Information Sheet," fact sheet, July 12, 2018.

United Nations, "Convention on Biodiversity," webpage, 2018. As of March 7, 2019:
http://www.un.org/en/events/biodiversityday/convention.shtml

University of Pittsburgh Institutional Review Board, "IRB Guidance: Privacy Versus Confidentiality," April 1, 2014. As of June 6, 2017:
http://www.irb.pitt.edu/sites/default/files/Privacy%20vs%20Conf_4.1.2014.pdf

Valantine, H., "The Science of Diversity and the Impact of Implicit Bias," National Institutes of Health, 2017. As of February 5, 2019:
https://diversity.nih.gov/sites/coswd/files/images/2017-12/implicit_bias_talk_for_toolkit_pdf_508c_0.pdf

van der Zande, Indira S. E., Rieke van der Graaf, Martijn A. Oudijk, and Johannes J. M. van Delden, "Vulnerability of Pregnant Women in Clinical Research," *Journal of Medical Ethics*, Vol. 43, No. 10, October 2017, pp. 657–663.

Weigmann, K., "The Ethics of Global Clinical Trials in Developing Countries, Participation in Clinical Trials Is Sometimes the Only Way to Access Medical Treatment. What Should Be Done to Avoid Exploitation of Disadvantaged Populations?" *Embo Reports*, Vol. 16, No. 5, 2015, pp. 566–570.

Wiesing, U., "The Declaration of Helsinki—Its History and Its Future," World Medical Association, November 11, 2014. As of March 7, 2019:
https://www.wma.net/wp-content/uploads/2017/01/Wiesing-DoH-Helsinki-20141111.pdf

WMA—*See* World Medical Association.

World Medical Association, *Medical Ethics Manual*, 2005. As of March 7, 2019:
https://www.wma.net/what-we-do/education/medical-ethics-manual/

———, *Annual Report*, 2017. As of March 7, 2019:
https://www.wma.net/wp-content/uploads/2018/04/WMA-Annual-Report-2017.pdf

———, "History—The Story of the WMA," 2018. As of March 7, 2019:
https://www.wma.net/who-we-are/history/

Worlock, K., "Access to the Literature: The Debate Continues," *Nature*, September 13, 2004.

Zook, M., S. Barocas, D. Boyd, K. Crawford, E. Keller, S. P. Gangadharan, A. Goodman, R. Hollander, B. A. Koenig, J. Metcalf, A. Narayanan, A. Nelson, and F. Pasquale, "Ten Simple Rules for Responsible Big Data Research," *Plos Computational Biology*, Vol. 13, No. 3, March 30, 2017, pp. 1–10. As of March 19, 2019:
https://journals.plos.org/ploscompbiol/article?id=10.1371/journal.pcbi.1005399